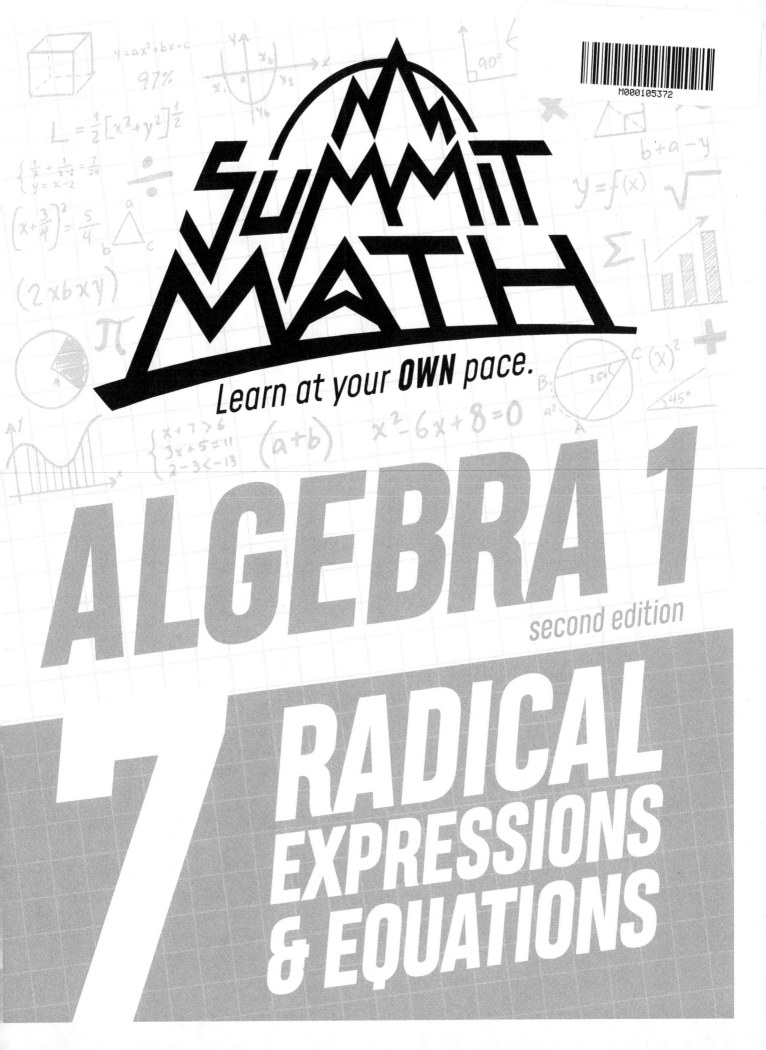

SUMMIT MATH

Learn at your **OWN** pace.

ALGEBRA 1

second edition

7 RADICAL EXPRESSIONS & EQUATIONS

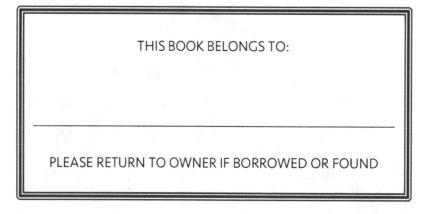

THIS BOOK BELONGS TO:

_____

PLEASE RETURN TO OWNER IF BORROWED OR FOUND

**DEDICATION**
To Lauren, Chloe, Dawson and Teagan

**ACKNOWLEDGEMENTS**
I started writing these books in 2013 to help my students learn better. I kept writing them because I received encouraging feedback from students, parents and teachers. Thank you to all who have used these books, pointed out my mistakes, and made suggestions along the way. Thank you to all of the students and parents who asked me to keep writing more books. Thank you to my family for supporting me through every step of this journey.

This book was typeset in the following fonts:
Seravek + Mohave + *Heading Pro*

Graphics in Summit Math books are made using the following resources:
Microsoft Excel | Microsoft Word | Desmos | Geogebra | Adobe Illustrator

First printed in 2017

Printed in the U.S.A.

Summit Math Books are written by Alex Joujan.

www.summitmathbooks.com

# INTRODUCTION

**Learning math through Guided Discovery:**
A Guided Discovery learning experience is designed to help you experience a feeling of discovery as you learn each new topic.

**Why this curriculum series is named Summit Math:**
Learning through Guided Discovery can be compared to climbing a mountain. Climbing and learning both require effort and persistence. In both activities, people naturally move at different paces, but they can reach the summit if they keep moving forward. Whether you race rapidly through these books or step slowly through each scenario, this curriculum is designed to keep advancing your learning until you reach the end of the book.

**Guided Discovery Scenarios:**
The Guided Discovery Scenarios in this book are written and arranged to show you that new math concepts are related to previous concepts you have already learned. Try to fully understand each scenario before moving on to the next one. To do this, try the scenario on your own first, check your answer when you finish, and then fix any mistakes, if needed. Making mistakes and struggling are essential parts of the learning process.

**Homework and Extra Practice Scenarios:**
After you complete the scenarios in each Guided Discovery section, you may think you know those topics well, but over time, you will forget what you have learned. Extra practice will help you develop better retention of each topic. Use the Homework and Extra Practice Scenarios to improve your understanding and to increase your ability to retain what you have learned.

**The Answer Key:**
The Answer Key is included to promote learning. When you finish a scenario, you can get immediate feedback. When the Answer Key is not enough to help you fully understand a scenario, you should try to get additional guidance from another student or a teacher.

**Star symbols:**
Scenarios marked with a star symbol ★ can be used to provide you with additional challenges. Star scenarios are like detours on a hiking trail. They take more time, but you may enjoy the experience. If you skip scenarios marked with a star, you will still learn the core concepts of the book.

**To learn more about Summit Math and to see more resources:**
Visit www.summitmathbooks.com.

As you complete scenarios in this part of the book, follow the steps below.

**Step 1: Try the scenario.**
Read through the scenario on your own or with other classmates. Examine the information carefully. Try to use what you already know to complete the scenario. Be willing to struggle.

**Step 2: Check the Answer Key.**
When you look at the Answer Key, it will help you see if you fully understand the math concepts involved in that scenario. It may teach you something new. It may show you that you need guidance from someone else.

**Step 3: Fix your mistakes, if needed.**
If there is something in the scenario that you do not fully understand, do something to help you understand it better. Go back through your work and try to find and fix your errors. Mistakes provide an opportunity to learn. If you need extra guidance, get help from another student or a teacher.

After Step 3, go to the next scenario and repeat this 3-step cycle.

### NEED EXTRA HELP?
### watch videos online

Teaching videos for every scenario in the Guided Discovery section of this book are available at www.summitmathbooks.com/algebra-1-videos.

# CONTENTS

Section 1
# EXPONENTS AND ROOTS

In this book, you will learn about radical expressions like the square root of 49, or $\sqrt{49}$. To understand square roots, you need to be familiar with exponents, which represent repeated multiplication. Let's go back to that topic to prepare your mind to learn about radical expressions.

1. Simplify the following expressions by raising each base to an exponent.

    a. $7^2$          b. $5^3$          c. $4^4$          d. $2^5$

2. When you learn about radical expressions, you learn how to think about exponents (or repeated multiplication) in reverse. Consider the following questions.

    a. What number when squared (raised to the 2nd power) is equal to 49?

    b. What number when cubed (raised to the 3rd power) is equal to 125?

    c. What number when raised to the 4th power is equal to 256?

    d. What number when raised to the 5th power is equal to 32?

The calculations you did in scenario 1 should have made it easy to answer the questions in scenario 2. In fact, the calculations in scenario 1 *must* be done in order to answer the questions in scenario 2.

3. At this point, we need to define some vocabulary.

    a. Since $7^2$ (or 7 squared) is 49, 7 is called the _____ of 49.

    b. Since $5^3$ (or 5 cubed) is 125, 5 is called the _____ of 125.

    c. Since $4^4$ (or 4 to the fourth power) is 256, 4 is called the _____ of 256.

    d. Since $2^5$ (or 2 to the fifth power) is 32, 2 is called the _____ of 32.

    e. If $A^n = B$, then $A$ is the _____ of $B$.

4. In mathematics, when there are many ways to show a concept, the most concise form will likely be used more often. In the table below, there are 3 equivalent (equal) forms of the same statement.

The 2nd Form is how you can interpret the question stated in the 1st Form.

The 3rd Form is the more concise version of the 1st Form.

| 1st Form | 2nd Form | 3rd Form |
|---|---|---|
| What is the square root of 49? | What number when squared is 49? | Simplify $\sqrt{49}$. |
| What is the cube root of 125? | What number when cubed is ____? | Simplify $\sqrt[3]{\phantom{x}}$. |
| What is the fourth root of 256? | What number raised to the 4th power is ____? | Simplify $\sqrt[4]{\phantom{x}}$. |
| What is the fifth root of 32? | What number raised to the 5th power is ____? | Simplify $\sqrt[5]{\phantom{x}}$. |
| What is the $n$th root of $B$? | What number raised to the $n$th power is ____? | Simplify $\sqrt[n]{\phantom{x}}$. |

5. Rewrite the following statements in a more concise form (the 3rd Form).

    a. What is the square root of 81?
    b. What is the cube root of 64?

    c. What is the fourth root of 81?
    d. What is the fifth root of 1?

6. Simplify the following expressions by raising each base to its specified exponent.

    a. $2^3$
    b. $3^3$
    c. $4^3$
    d. $5^3$

7. Use your work in the previous scenario to simplify the following expressions.

    a. $\sqrt[3]{8}$
    b. $\sqrt[3]{27}$
    c. $\sqrt[3]{64}$
    d. $\sqrt[3]{125}$

8. Simplify the following expressions by raising each base to its specified exponent.

    a. $2^4$
    b. $3^4$
    c. $4^4$
    d. $5^4$

9. Use your work in the previous scenario to simplify the following expressions.

    a. $\sqrt[4]{16}$
    b. $\sqrt[4]{81}$
    c. $\sqrt[4]{256}$
    d. $\sqrt[4]{625}$

10. Use what you have learned so far to answer each question below.

    a. What is the square root of 81?        b. What is the cube root of 64?

    c. What is the fourth root of 81?        d. What is the fifth root of 1?

A root is the inverse operation for an exponent because the purpose of a root is to <u>undo</u> the work of an exponent. Thus, to find a root, you need to think about exponents. For example, consider the sixth root of 729, or $\sqrt[6]{729}$. To find that root, you need to realize that $3^6 = 729$. Similarly, there is no quick way to simplify $\sqrt[3]{343}$, unless you have already memorized that $7^3 = 343$. Even a calculation like finding the square root of 25, $\sqrt{25}$, is only possible if you know that $5^2 = 25$.

11. Compute the following roots.

    a. $\sqrt[3]{27}$        b. $\sqrt[4]{16}$        c. $\sqrt[5]{1}$        d. $\sqrt{0}$

12. Compute the following roots.

    a. $\sqrt[4]{256}$        b. $\sqrt[3]{-8}$        c. $\sqrt[5]{-1024}$        d. $\sqrt{-16}$

13. In the previous scenario, why is it that you cannot compute $\sqrt{-16}$ ?

14. Apply what you know about fractions to simplify the following expressions.

    a. $\left(\dfrac{2}{3}\right)^2$        b. $\left(\dfrac{2}{3}\right)^3$        c. $\left(-\dfrac{2}{3}\right)^2$        d. $\left(-\dfrac{2}{3}\right)^3$

15. Simplify the following expressions (apply the exponent to remove the parentheses).

    a. $\left(-\dfrac{3}{5}\right)^3$        b. $\left(-\dfrac{x^2}{10}\right)^4$        ★c. $\left(-\dfrac{6x}{11}\right)^2$        ★d. $\left(-\dfrac{2}{xy^3}\right)^5$

16. In the previous scenario, $\left(-\dfrac{3}{5}\right)^3$ is simplified to become $-\dfrac{27}{125}$. It follows that the cube root of $-\dfrac{27}{125}$, written as $\sqrt[3]{-\dfrac{27}{125}}$, must be $-\dfrac{3}{5}$. What is the cube root of $-\dfrac{8}{27}$?

17. Compute the following roots.

     a. $\sqrt[4]{\dfrac{16}{81}}$      b. $\sqrt[3]{-\dfrac{1}{8}}$      c. $\sqrt[5]{-\dfrac{1024}{3125}}$      d. $\sqrt{-\dfrac{4}{9}}$

18. In the previous scenario, why is it that you cannot compute $\sqrt{-\dfrac{4}{9}}$?

19. Determine the square root of 10.

At this point, you face a challenge. To determine the square root of 10, you must run through your list of perfect squares (1, 4, 9, 16, ...), but 10 is not in your list. Since 10 is not in your list, you can only determine the square root of 10 by making a guess or by using a calculator.

20. ★As you have seen, the list of integers that are called perfect squares is limited. How many integers from 1 to 100, inclusive, are not perfect squares?

21. ★What is the probability that an integer will be a perfect square if you randomly select an integer from the following range?

     a. 1 to 20      b. 1 to 42      c. 1 to 72

Since the probability that an integer is a perfect square is low, you need to learn how to figure out the square roots of numbers that are not perfect squares.

# NOTES

Use this page to record important ideas in the previous section or
for any other writing that helps you learn the topics in this book.

8

## Section 2
# COMPARING RATIONAL AND IRRATIONAL NUMBERS

Consider the square root of 10 again. Logically, since the square root of 9 is 3 and the square root of 16 is 4, then the value of the square root of 10 must be between 3 and 4. You may not know its exact value, but you can take a guess.

22. Without a calculator, use the roots that you know to <u>guess</u> the value of each root below. Round your guess to the nearest tenth.

      a. $\sqrt{37}$            b. $\sqrt{60}$            c. $\sqrt{98}$

23. Although you should be able to approximate the value of many square roots by simply figuring out which perfect square they are close to, the quickest way to accurately determine the value of a square root is to use a calculator. Take a moment to figure out how to determine the following square roots with a calculator. Round your answers to three decimal places.

      a. $\sqrt{37}$            b. $\sqrt{60}$            c. $\sqrt{98}$

24. When you use a calculator to compute the square root of 37, is the entire number visible on the calculator screen?

A calculator is a great tool for quickly finding the value of a specific square root, but you usually cannot write down the exact value of a square root. The decimals just keep going. Mathematicians have a name for this type of numbers: <u>irrational</u>. The next collection of scenarios will explain the words "rational" and "irrational."

25. A number that can be expressed as a ratio of two integers (a fraction) is called a rational number. You don't have to look too hard to see ratio in the word rational.

      Are integers rational numbers? How can you support your claim?

If you could take all of the integers on the number line and pull them off, the numbers that remained on the number line would all contain decimals, such as 0.5, 2.18, $-5.\overline{3}$, etc... These numbers that contain decimals can be split into three groups.

      Group 1. Decimals that terminate (end).
      Group 2. Decimals that have a repeating pattern.
      Group 3. Decimals that neither terminate nor repeat.

26. **Group 1: Numbers with terminating decimals.** Every number with a terminating decimal can be written as a fraction. Therefore, these types of numbers are rational. Some terminating decimals are shown below. Write them in fraction form. You may use a calculator to help you.

    a. 0.5               b. 0.25             c. 0.2             d. 0.75

27. **Group 2. Numbers with repeating decimals.** Every number with a repeating decimal can be written as a fraction. Therefore, these types of numbers are also rational. Some repeating decimals are shown below. Write them in fraction form. You may use a calculator to help you.

    a. $0.\overline{3}$             b. $0.\overline{1}$             c. $0.\overline{6}$             d. $0.\overline{5}$

28. **Group 3. Numbers that have decimals that neither terminate nor repeat.** These have the unique quality that they cannot be written as fractions. Therefore, these types of numbers are <u>not</u> rational. These numbers are called **ir**rational.

    a. The most famous irrational number is probably $\pi$. What is the approximate value of $\pi$?

    b. In an earlier scenario, you used a calculator to determine the square root of $37 \rightarrow \sqrt{37}$.

    What digits are shown on your calculator when you compute $\sqrt{37}$ ?

The digits for the decimal form of $\sqrt{37}$ keep going until the calculator screen ends, which gives the impression that the decimal value of the number ends at the end of the calculator screen. If the digits truly end, then $\sqrt{37}$ is <u>rational</u>, but the digits after the decimal point in $\sqrt{37}$ actually keep going without ending. The digits do not form a repeating pattern so this number is called **ir**rational.

29. Consider the numbers shown below. Circle all of the numbers that are **ir**rational.

$$\sqrt{1} \qquad \sqrt{3} \qquad \sqrt{4} \qquad \sqrt{8} \qquad \sqrt{9} \qquad \sqrt{15} \qquad \sqrt{16}$$

30. Consider the first 100 square roots $\left(\sqrt{1}, \sqrt{2}, \sqrt{3}, \dots, \sqrt{100}\right)$. How many are **ir**rational?

Since most of the roots from $\sqrt{1}$ up to $\sqrt{100}$ (and well beyond) are irrational, you will need to become comfortable with irrational numbers. Integers and fractions have their place, but it is time to expand your mathematical awareness.

# NOTES

Use this page to record important ideas in the previous section or
for any other writing that helps you learn the topics in this book.

12

*Section 3*
# SIMPLIFYING RADICAL EXPRESSIONS

The next scenarios use your familiarity with rational square roots to help you learn how to simplify square roots that are irrational.

31. Simplify each product as much as possible.  Analyze the results as you work through each column.

   a.  $\sqrt{4 \cdot 9}$        c.  $\sqrt{25 \cdot 4}$        e.  $\sqrt{16 \cdot 25}$        g.  $\sqrt{16 \cdot 9}$

   b.  $\sqrt{4} \cdot \sqrt{9}$        d.  $\sqrt{25} \cdot \sqrt{4}$        f.  $\sqrt{16} \cdot \sqrt{25}$        h.  $\sqrt{16} \cdot \sqrt{9}$

32. Simplify each product as much as possible.  Analyze the results as you work through each column.

   a.  $\sqrt{16 \cdot 4}$        c.  $\sqrt{100 \cdot 49}$        e.  $\sqrt{36 \cdot 64}$        g.  $\sqrt{81 \cdot 25}$

   b.  $\sqrt{16} \cdot \sqrt{4}$        d.  $\sqrt{100} \cdot \sqrt{49}$        f.  $\sqrt{36} \cdot \sqrt{64}$        h.  $\sqrt{81} \cdot \sqrt{25}$

33. Your work in the previous scenario should suggest the following properties:

   a.  $\sqrt{A \cdot B}$ is equivalent to ...        b.  $\sqrt{X} \cdot \sqrt{Y}$ is equivalent to ....

34. Now consider a product such as $\sqrt{25} \cdot \sqrt{3}$ . You can simplify $\sqrt{25}$ as 5, but you cannot rewrite $\sqrt{3}$ . As a result, the most simplified form of $\sqrt{25} \cdot \sqrt{3}$ is _____.

35. What is the most simplified form of $\sqrt{4} \cdot \sqrt{5}$ ?

36. The expression $2\sqrt{5}$ means "2 multiplied by the square root of 5." Another way to say this is "2 times the square root of 5." In order to make it easier to say, you can shorten this to "two root five." The expression $4\sqrt{11}$ can be written out with words as "four root eleven." Write out each of the following expressions using only words.

   a. $5\sqrt{3}$          b. $7\sqrt{6}$          c. $4\sqrt{10}$

37. Simplify each product as much as possible.

   a. $\sqrt{4}\cdot\sqrt{3}$     b. $\sqrt{9}\cdot\sqrt{2}$     c. $\sqrt{2}\cdot\sqrt{25}$     d. $\sqrt{5}\cdot\sqrt{2}$

38. In the previous scenario, $2\sqrt{3}$ is the most simplified form of $\sqrt{4}\cdot\sqrt{3}$. You can use the property that $\sqrt{A\cdot B}=\sqrt{A}\cdot\sqrt{B}$ to rewrite $\sqrt{4}\cdot\sqrt{3}$ as $\sqrt{4\cdot3}$ which is also $\sqrt{12}$. This reveals that the most simplified form of $\sqrt{12}$ is $2\sqrt{3}$.

   a. What is the most simplified form of $\sqrt{27}$ ?

   b. Another way to state the previous result is that the square root of 27 has the same numerical value as 3 multiplied by the square root of _____.

   c. Since $\sqrt{12}$ is the same as $2\sqrt{3}$, it follows that $\sqrt{12}$ is ____ times larger than $\sqrt{3}$.

39. Radical expressions are simplified below. Fill in the missing values in each scenario.

   a. $\sqrt{18}$          b. $\sqrt{48}$          c. $\sqrt{20}$
      $\sqrt{9\cdot2}$         $\sqrt{\underline{\quad}\cdot3}$         $\sqrt{\underline{\quad}\cdot\underline{\quad}}$
      $\sqrt{9}\sqrt{2}$         $\sqrt{\underline{\quad}}\sqrt{3}$         $\sqrt{\underline{\quad}}\sqrt{\underline{\quad}}$
      $\underline{\quad}\sqrt{2}$         $\underline{\quad}\sqrt{3}$         $\underline{\quad}\sqrt{\underline{\quad}}$

40. Write down the first ten perfect squares.

41. What is the largest number that is a factor of the number shown and also a perfect square?

      a. 32             b. 45             c. 500            d. 192

42. For the number shown, find all of the factors of that number that are perfect squares.

      a. 32             b. 80             c. 36            d. 288

43. Simplify each expression shown by finding the largest perfect square factor.

      a. $\sqrt{20}$            b. $\sqrt{24}$            c. $\sqrt{28}$

44. Simplify each expression shown by finding the largest perfect square factor.

      a. $\sqrt{45}$            b. $\sqrt{54}$            c. $\sqrt{63}$

45. Simplify each expression shown by finding the largest perfect square factor.

      a. $\sqrt{50}$            b. $\sqrt{75}$            c. $\sqrt{22}$

46. Simplify each expression. Write your answer in the form $A\sqrt{B}$.

    a.  $\sqrt{98}$                                 b.  $\sqrt{40}$

47. Simplify each expression. Write your answer in the form $A\sqrt{B}$.

    a.  $\sqrt{162}$                               b.  $\sqrt{104}$

48. If the square root of a number is $2\sqrt{2}$, what is the number?

49. If the square root of a number is $3\sqrt{11}$, what is the number?

50. Some expressions may look simplified, but they contain expressions that still contain perfect square factors. Simplify each of the following expressions completely.

    a.  $2\sqrt{8}$                               b.  $3\sqrt{12}$

51. Simplify each expression completely.

    a.  $7\sqrt{32}$             b.  $5\sqrt{9}$             c.  $2\sqrt{14}$

# NOTES

Use this page to record important ideas in the previous section or
for any other writing that helps you learn the topics in this book.

Section 4
# SOLVING EQUATIONS WITH RADICALS

52. Determine the value of $x$ that makes each equation true.

       a. $\sqrt{x}=1$          b. $\sqrt{x}=5$          c. $\sqrt{x}=2\sqrt{2}$

53. If $\sqrt{x}=H$, then $x=$ _____.

54. If $\sqrt{y+1}=L$, then $y+1=$ _____.

55. When you solve an equation, you undo operations one at a time to isolate a variable.

    a. To solve $x+3.1=7.4$, you can undo "+3.1" by _____ on both sides of the equation.

    b. To solve $\dfrac{x}{3}=1.5$, you can undo "dividing by 3" by _____ on both sides.

    c. What operation will undo a square root? In other words, if you take the square root of a number, how can you undo that operation to get back to the original number?

56. It may help to use specific numbers. If you compute the square root of 100, the result is 10. What operation can you perform to make the 10 become 100 again?

57. If $\sqrt{z-2}=P$, then $z-2=$ _____ and $z=$ _____

58. Determine the value of $x$ that makes each equation true.

       a. $\sqrt{x-1}=4$         b. $\sqrt{x+4}=2$         c. $\sqrt{2x-3}=1$

59. If $\sqrt{x}=-2$, then $x=$ _____.

In the previous scenario, the equation has no solution, but this may be confusing. If you replace $x$ with 4, the equation becomes $\sqrt{4}=-2$, or $2 = -2$, which is a false statement.

60. Determine the value of $x$ that makes each equation true.

    a. $\sqrt{x-1}=10$      b. $\sqrt{x-1}=\sqrt{5}$      ★c. $\sqrt{x+4}=2\sqrt{3}$      ★d. $\sqrt{2x-3}=3\sqrt{5}$

61. Solve each equation. Notice how the structure of these equations is slightly different than the equations in the previous scenario.

    a. $2+\sqrt{x-2}=7$      b. $3+\sqrt{1-x}=5$      ★c. $5+\sqrt{1+3x}=1$

62. Simplify each expression.

    a. $(x+2)^2$      b. $(x-1)^2$      c. $(x+5)^2$      d. $(3x-7)^2$

63. Solve each equation.

    a. $x^2+5x+6=0$      b. $x^2-3x-18=0$      ★c. $2x^2-x-6=0$

64. The following equation is partially solved for you. Fill in the blank shown and then keep going until you have solved the equation.

a. $\sqrt{2x}=x-4 \rightarrow \left(\sqrt{2x}\right)^2=(x-4)^2$

$$2x=x^2-8x+16$$
$$-2x \qquad -2x$$
$$0=\underline{\hspace{3cm}}$$

b. $\sqrt{-3x+1}=x+3 \rightarrow \left(\sqrt{-3x+1}\right)^2=(x+3)^2$

$$-3x+1=\underline{\hspace{3cm}}$$

65. Use what you did in the previous scenario to solve each equation.

a. $\sqrt{x+3}=x+1$

b. $\sqrt{x-3}=x-3$

66. In the previous scenario, you may have assumed that both solutions made the original equation true. If you have not already done this, check your solutions by plugging them back into the original equation. Confirm that each solution makes the original equation a true statement.

67. ★Solve the equation.

$$\sqrt{7-2x}=x+4$$

When you square both sides in the previous scenario, the right side of the equation becomes a trinomial. After you bring all of the terms to one side of the equation, you can factor the expression. Your two separate factors result in two separate solutions (unless the factors are identical).

68. Solve each equation.

    a. $\sqrt{x+1}=x-5$

    b. $\sqrt{4x+8}=x+3$

69. ★Solve the equation.

    $$\sqrt{21-5x}=-5+x$$

70. When a number, $N$, is decreased by 11, the square root of the result is 5. What is the value of $N$?

# NOTES

Use this page to record important ideas in the previous section or
for any other writing that helps you learn the topics in this book.

*Section 5*

# MULTIPLICATION WITH RADICALS

71. Apply what you have learned so far to simplify the following expressions.

   a. $\sqrt{6}\cdot\sqrt{2}$
   b. $\sqrt{5}\cdot\sqrt{5}$
   c. $\left(\sqrt{10}\right)^2$
   d. $\sqrt{8}\cdot\sqrt{14}$

72. Simplify each expression.

   a. $2(3\cdot4)$
   b. $2(3x)$
   c. $2\left(3\sqrt{2}\right)$
   d. $2\left(A\sqrt{B}\right)$

73. Simplify each expression. Write your answer in the form $A\sqrt{B}$, if possible.

   a. $5\left(4\sqrt{3}\right)$
   b. $3\left(2\sqrt{4}\right)$
   c. $2\left(3\sqrt{8}\right)$

74. Simplify each expression. Write your answer in the form $A\sqrt{B}$, if possible.

   a. $4\sqrt{27}$
   ★b. $\sqrt{8}\left(2\sqrt{2}\right)$

75. Multiply each group of expressions and simplify the result as much as you can.

   a. $2(4)\cdot2(3)$
   b. $5x\cdot3y$
   c. $5\sqrt{2}\cdot3\sqrt{3}$

76. Multiply each group of expressions and simplify the result as much as you can.

   a. $11\sqrt{7}\cdot4\sqrt{2}$
   b. $A\sqrt{B}\cdot C\sqrt{D}$
   c. $\left(2\sqrt{2}\right)^2$

77. Multiply each group of expressions and simplify the result as much as you can. Write your answer in the form $A\sqrt{B}$, if possible.

    a. $2\sqrt{5} \cdot 4\sqrt{10}$
    b. $\left(-2\sqrt{15}\right)\left(3\sqrt{5}\right)$
    c. $\left(2\sqrt{6}\right)\left(-3\sqrt{6}\right)$
    d. $\left(-3\sqrt{7}\right)^2$

78. ★Do not use a calculator. If $\sqrt{2}$ is 1.41 when rounded to 2 decimal places, what is the value of $\sqrt{18}$ when rounded to 2 decimal places?

79. ★If $\sqrt{3}$ is 1.73 when rounded to 2 decimal places, what is the value of $\sqrt{75}$ when rounded to 2 decimal places? Do not use a calculator.

80. What value of x makes each equation true?

    a. $\sqrt{x+2}=3$
    b. $5+\sqrt{x+2}=10$
    c. $4-\sqrt{x+2}=3$

81. If $\sqrt{G}=2\sqrt{6}$, what is the value of G? If $\sqrt{H}=4\sqrt{3}$, what is the value of H?

82. Without a calculator, find a way to arrange the following numbers from smallest to largest.

    a. $2\sqrt{5},\ 3\sqrt{2},\ \sqrt{19}$
    b. $7\sqrt{2},\ 4\sqrt{7},\ 3\sqrt{10},\ 6\sqrt{3}$

At this point, you have learned how to solve equations that contain radical expressions. You have also discovered properties that allow you to simplify radicals and multiply radicals. The next set of properties to consider are those that involve addition and subtraction.

# NOTES

Use this page to record important ideas in the previous section or
for any other writing that helps you learn the topics in this book.

Section 6

# ADDITION AND SUBTRACTION WITH RADICALS

83. Simplify each sum or difference as much as possible.

   a. $\sqrt{1+4}$

   b. $\sqrt{1}+\sqrt{4}$

   c. $\sqrt{9-1}$

   d. $\sqrt{9}-\sqrt{1}$

   e. $\sqrt{4+12}$

   f. $\sqrt{4}+\sqrt{12}$

84. Is the expression $\sqrt{1}+\sqrt{1}$ equal to the expression $\sqrt{1+1}$?

85. Simplify each sum or difference as much as possible.

   a. $\sqrt{9+16}$

   b. $\sqrt{9}+\sqrt{16}$

   c. $\sqrt{100-25}$

   d. $\sqrt{100}-\sqrt{25}$

   e. $\sqrt{100-64}$

   f. $\sqrt{100}-\sqrt{64}$

86. Determine if the following properties exist.

   a. $\sqrt{A}+\sqrt{B}=\sqrt{A+B}$

   b. $\sqrt{A}-\sqrt{B}=\sqrt{A-B}$

87. At this point you have seen that $\sqrt{2}\cdot\sqrt{3}$ is equivalent to $\sqrt{2\cdot3}$, but your work above reveals that $\sqrt{2}+\sqrt{3}$ does not equal $\sqrt{2+3}$. In mathematics, there are many instances where terms can be multiplied together, but they cannot also be added together.

   a. For example, $3x+2x$ is $5x$, but $3x+2y$ cannot be combined. What is the simplified form of $3x\cdot2y$?

   b. The expression $y+3y+7y$ can be simplified, but $4M+2N$ cannot be combined. What is the simplified form of $4M\cdot2N$?

   c. The expression $xy^4+3xy^4+5xy^4$ can be simplified, but $5xy^2+6x^2y$ cannot. What is the simplified form of $5xy^2\cdot6x^2y$?

88. Explain why each statement below is false.

    a. $\sqrt{4}+\sqrt{4}=\sqrt{8}$
    b. $\sqrt{4}+\sqrt{9}=\sqrt{13}$
    c. $\sqrt{36}-\sqrt{25}=\sqrt{11}$

89. Simplify each expression as much as you can.

    a. 2 eggs plus 3 eggs
    b. $2x+3x$
    c. $2x^2+3x^2$
    d. $2\sqrt{2}+3\sqrt{2}$

90. Simplify each expression as much as you can.

    a. $1\sqrt{2}+1\sqrt{2}$
    b. $\sqrt{3}+\sqrt{3}$
    c. $\sqrt{5}+2\sqrt{5}$

91. Simplify each expression as much as you can.

    a. $6\sqrt{2}-4\sqrt{2}$
    b. $10\sqrt{3}-7\sqrt{3}$
    c. $2\sqrt{5}-6\sqrt{5}$

92. What is the simplified form of $\sqrt{5}+\sqrt{6}$ ?

93. Explain how you can determine if two radical expressions are like terms.

94. The expressions $\sqrt{20}$ and $\sqrt{45}$ cannot be combined in their current forms. However, if you simplify each radical expression as much as possible, it becomes clear that they can be combined. In a different form, the numbers become like terms. What is the simplified form of $\sqrt{20}+\sqrt{45}$ ?

95. Simplify the expression $\sqrt{75} - \sqrt{48}$ .

96. Simplify each expression.

    a. $\sqrt{8} - 2\sqrt{2}$          b. $-\sqrt{5} + \sqrt{80}$          c. $\sqrt{12} - 2\sqrt{3}$

97. Simplify each expression as much as possible.

    a. $\sqrt{2} + \sqrt{8}$          b. $-\sqrt{10} + \sqrt{40}$

98. Simplify each expression as much as possible.

    a. $\sqrt{80} - \sqrt{20}$          b. $\sqrt{24} - \sqrt{54}$          c. $\sqrt{18} + \sqrt{27}$

99. ★Simplify each expression as much as possible.

    a. $\sqrt{72} + \sqrt{8} - \sqrt{128}$          b. $4\sqrt{20} + 3\sqrt{75} - 2\sqrt{125} - \sqrt{108}$

© Alex Joujan, 2020

# NOTES

Use this page to record important ideas in the previous section or
for any other writing that helps you learn the topics in this book.

Section 7

# THE DISTRIBUTIVE PROPERTY WITH RADICALS

100. Use the Distributive Property to simplify the following expressions.

    a. $3(x+5)$        b. $w(w^2-11)$        c. $-2y^3(y^2-7y+1)$

101. Use what you know about the Distributive Property to simplify the following expressions.

    a. $3(2+\sqrt{3})$        b. $\sqrt{6}(\sqrt{2}-4)$        c. $\sqrt{5}(1-\sqrt{10})$

102. Simplify the following expressions.

    a. $(x-2)(x+3)$      b. $(x-2)^2$      c. $(2x+5)(2x-5)$      d. $(3x+1)^2$

103. Simplify the following expressions.

    a. $(1+\sqrt{2})^2$        b. $(4-\sqrt{3})^2$        c. $(3+2\sqrt{5})^2$

104. Simplify the following expressions.

    a. $(3+\sqrt{7})(3-\sqrt{7})$      b. $(1+\sqrt{5})(1-\sqrt{5})$      c. $(5+2\sqrt{6})(5-2\sqrt{6})$

105. As a matter of review, if an expression is *completely* simplified and it *still* contains a radical expression, such as $\sqrt{10}$ or $3-\sqrt{2}$, the expression is identified as irrational. In the previous scenario, the expressions are rational. Why are they rational?

106. Which of the following expressions are rational? If it is helpful, simplify each expression.

a. $\left(2-\sqrt{5}\right)\left(2+\sqrt{5}\right)$

b. $\left(1-\sqrt{3}\right)^2$

c. $\left(3+2\sqrt{5}\right)\left(3-2\sqrt{5}\right)$

107. Which of the following expressions are rational? If it is helpful, simplify each expression.

a. $\left(\sqrt{2}+5\right)^2$

b. $\left(3-2\sqrt{7}\right)\left(3+2\sqrt{7}\right)$

c. $\left(3-\sqrt{3}\right)^2$

108. Fill in the blank. Identify an expression that will make each product rational.

a. $\sqrt{2}\cdot$____

b. $3\sqrt{3}\cdot$____

c. $\left(5-\sqrt{3}\right)($_____$)$

d. $\left(1+3\sqrt{2}\right)($_____$)$

109. Numbers like 9, 16, and 25 can be referred to as perfect squares because $3^2=9$, $4^2=16$, and $5^2=25$. Are perfect squares always integers?

110. A square has an area of 18 in². What is the perimeter of this square?

Use this page to record important ideas in the previous section or
for any other writing that helps you learn the topics in this book.

Section 8
# SIMPLIFYING RADICALS WITH FRACTIONS

38

111. Use what you have learned so far to simplify each root as much as possible.

   a. $\sqrt{\dfrac{1}{4}}$
   b. $\sqrt{\dfrac{4}{9}}$
   c. $\sqrt{\dfrac{25}{9}}$
   d. $\sqrt{\dfrac{25}{100}}$

112. Simplify each product as much as possible.

   a. $\sqrt{\dfrac{5}{3}}\cdot\sqrt{\dfrac{3}{5}}$
   b. $\sqrt{9}\cdot\sqrt{\dfrac{1}{9}}$
   c. $\sqrt{\dfrac{4}{3}}\cdot\sqrt{3}$
   d. $\sqrt{\dfrac{28}{5}}\cdot\sqrt{\dfrac{45}{7}}$

113. Working with fractions leads naturally to the concept of division. Find out if any patterns emerge when division is combined with square roots in the expressions below. Simplify each quotient as much as possible and analyze the results.

   a. $\dfrac{\sqrt{16}}{\sqrt{4}}$
   b. $\dfrac{\sqrt{81}}{\sqrt{9}}$

   c. $\sqrt{\dfrac{16}{4}}$
   d. $\sqrt{\dfrac{81}{9}}$

114. Simplify each expression and analyze the results.

   a. $\dfrac{\sqrt{20}}{\sqrt{4}}$
   b. $\dfrac{\sqrt{54}}{\sqrt{9}}$

   c. $\sqrt{\dfrac{20}{4}}$
   d. $\sqrt{\dfrac{54}{9}}$

115. Your work in the previous scenario should suggest the following property:

a. $\sqrt{\dfrac{A}{B}}$ is equivalent to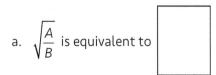

b. $\dfrac{\sqrt{X}}{\sqrt{Y}}$ is equivalent to ☐

116. Simplify each quotient as much as possible.

a. $\sqrt{\dfrac{5}{4}}$

b. $\dfrac{\sqrt{2}}{\sqrt{9}}$

c. $\dfrac{4\sqrt{12}}{\sqrt{25}}$

117. This may seem repetitive, but once again, simplify each quotient as much as possible.

a. $\sqrt{\dfrac{4}{32}}$

b. $\dfrac{\sqrt{9}}{\sqrt{5}}$

c. $\dfrac{6\sqrt{8}}{\sqrt{3}}$

118. After you have checked your results, look carefully at your simplified fractions in the previous two scenarios. Can you identify a structural difference between the fractions in those two scenarios?

Sometimes, when you are working with radical expressions, a radical expression will show up in the denominator of a fraction. This makes the denominator irrational. It is typical to rewrite the fraction to make the denominator a rational number. This is referred to as *rationalizing* the denominator. For example, the expression $\dfrac{1}{\sqrt{2}}$ can be rewritten as $\dfrac{\sqrt{2}}{2}$. How? The next group of scenarios reveals this.

# NOTES

Use this page to record important ideas in the previous section or
for any other writing that helps you learn the topics in this book.

## Section 9
# HOW TO RATIONALIZE A DENOMINATOR

One concept in mathematics that appears throughout the topics you *have* learned and *will* learn is that of multiplying by 1. The number system is defined so that when you multiply an expression by 1, the value of the expression <u>does not change</u>.

119. For example, $3 \cdot 1$ is ___, $y \cdot 1$ is still ___, and $(x+3) \cdot 1$ is just _____.

Since an expression does not lose its original identity (its value) when multiplied by 1, this is called the Identity Property of Multiplication. This property may not seem useful if we only multiply by the actual number "1". It becomes more complex – and much more useful – when you consider that "1" can take on many forms.

120. For example, $\dfrac{3}{3}$, $\dfrac{x+1}{x+1}$ and $\dfrac{\sqrt{2}}{\sqrt{2}}$ are each equivalent to the number ____.

Notice how multiplying by "1" allows you to add fractions in 3 separate scenarios below.

121. Adding $\dfrac{1}{2}+\dfrac{1}{6}$ is difficult until you multiply $\dfrac{1}{2} \cdot \dfrac{3}{3}$ and realize that $\dfrac{1}{2}$ can be written as $\dfrac{3}{6}$. Now it becomes clear that $\dfrac{1}{2}+\dfrac{1}{6}=\dfrac{}{6}+\dfrac{}{6}=$ _____.

122. Consider $2-\dfrac{3}{x}$. It does not look like two fractions until you write it as $\dfrac{2}{1}-\dfrac{3}{x}$. When you multiply $\dfrac{2}{1}$ by the expression $\dfrac{x}{x}$, you can rewrite 2 as $\dfrac{2}{1} \cdot \dfrac{x}{x}$ or $\dfrac{2x}{x}$. Now you can conclude that the expression

$2-\dfrac{3}{x}=\dfrac{}{x}-\dfrac{}{x}=\dfrac{}{x}$.

123. Finally, suppose you try to add $\dfrac{3\sqrt{2}}{2}$ and $\dfrac{1}{\sqrt{2}}$. If you multiply $\dfrac{1}{\sqrt{2}}$ by $\dfrac{\sqrt{2}}{\sqrt{2}}$, it becomes $\dfrac{\sqrt{2}}{\sqrt{4}}$, or $\dfrac{\sqrt{2}}{2}$ when simplified. Now you can conclude that $\dfrac{3\sqrt{2}}{2}+\dfrac{1}{\sqrt{2}}=\dfrac{}{2}+\dfrac{}{2}=\dfrac{}{2}$.

124. If you were to multiply each expression below by a form of 1 to make the denominator a rational number, what form of 1 would you use?  Each expression will require a different form of 1.

    a. $\dfrac{3}{\sqrt{2}}$          b. $\dfrac{1}{\sqrt{6}}$          c. $\dfrac{5}{\sqrt{3}}$          d. $\dfrac{7}{\sqrt{5}}$

125. Multiply each expression by a form of 1 that makes the denominator a rational number.  Simplify the result as much as you can.

    a. $\dfrac{4}{\sqrt{3}}$          b. $\dfrac{5}{\sqrt{5}}$          c. $\dfrac{14}{\sqrt{7}}$          ★d. $\dfrac{8\sqrt{3}}{\sqrt{2}}$

126. Consider the expression $\dfrac{2}{3\sqrt{3}}$.

    a.  Multiply the expression by $\dfrac{3\sqrt{3}}{3\sqrt{3}}$ and simplify.

    b.  Now multiply the original expression by $\dfrac{\sqrt{3}}{\sqrt{3}}$ and simplify again.

    c.  What do you notice?

127. Multiply each expression by a form of 1 that makes the denominator a rational number.  Simplify the result as much as you can.

    a. $\dfrac{6}{\sqrt{11}}$          b. $\dfrac{10}{\sqrt{2}}$          c. $\dfrac{6}{5\sqrt{2}}$

128. Simplify each expression. Make the denominator rational in your final result.

a. $\sqrt{\dfrac{16}{3}}$

b. $\sqrt{\dfrac{1}{2}}$

★c. $\sqrt{\dfrac{9}{20}}$

129. Add the following fractions. Rationalize the denominator in your final answer.

a. $\dfrac{2\sqrt{3}}{5}+\dfrac{3\sqrt{3}}{2}$

b. $\dfrac{2}{\sqrt{3}}+\dfrac{5\sqrt{3}}{3}$

130. Add the following fractions. Rationalize the denominator in your final answer.

a. $\dfrac{1}{\sqrt{3}}+\dfrac{1}{\sqrt{2}}$

★b. $\sqrt{\dfrac{4}{5}}+\sqrt{\dfrac{1}{3}}$

# NOTES

Use this page to record important ideas in the previous section or
for any other writing that helps you learn the topics in this book.

Section 10
# THE PYTHAGOREAN THEOREM

131. You have done a lot of work with square roots. Let's take a brief break from all of that. Pull out a ruler, and determine the missing side lengths for each right triangle below.

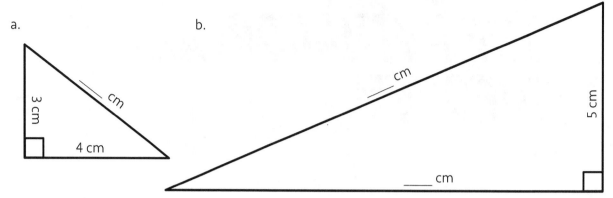

a.

b.

132. Around 2400 years ago, a man named Pythagoras of Samos devoted his life to studying philosophy, science, and most famously, mathematics. Although his contributions to the field of mathematics are somewhat unclear, his work with right triangles apparently led to the discovery that there exists a rather simple relationship between the 3 side lengths of every right triangle. Since he is credited with this discovery, it is known as the Pythagorean Theorem. Write the Pythagorean Theorem below.

133. Using the Pythagorean Theorem involves squaring expressions and simplifying square roots. Simplify each expression below as much as you can.

a. $\sqrt{6^2+8^2}$

b. $\sqrt{3^2+1^2}$

c. $\sqrt{2^2+6^2}$

d. $\sqrt{\left(\sqrt{11}\right)^2+5^2}$

134. Solve each equation.

a. $x^2+4^2=6^2$

b. $5^2+12^2=x^2$

c. $\left(2\sqrt{3}\right)^2+x^2=4^2$

135. The Pythagorean Theorem has a simple structure of $a^2+b^2=c^2$ but its simplicity can cause confusion. If the side lengths are labeled $a$, $b$, and $c$, the longest side must be labeled as $c$. The other two side lengths, known as the <u>legs</u> of the right triangle, are interchangeable as $a$ and $b$ or $b$ and $a$. Explain why it doesn't matter which leg is $a$ and which is $b$.

136. Which sides are called the <u>legs</u> of the right triangle?

a.

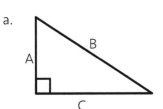

b.

c.

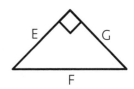

137. What is the name for the longest side of a right triangle?

138. Use the Pythagorean Theorem to determine the missing side length for each triangle shown.  The triangles below are NOT drawn to scale.

a.

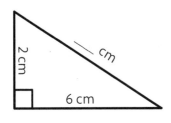

b.

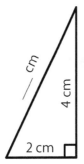

c.

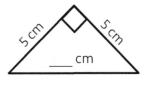

139. ★Determine the missing side length for the triangle shown.

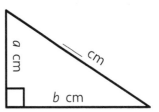

140. Determine the missing side length in each triangle.  Write the length in two forms: 1) in simplest radical form, and 2) rounded to the nearest tenth.  Triangles are not drawn to scale.

a.

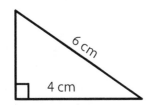

b.

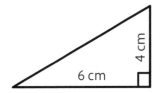

141. Determine the missing side length for each triangle shown.  Triangles are not drawn to scale.

a.

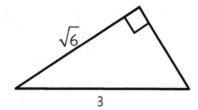

b.

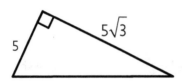

c.

142. A right triangle has side lengths of 5, 10, and *M*.  What is the value of *M* if 5 < *M* < 10?

143. ★Draw 2 unique right triangles such that each triangle has a hypotenuse of $\sqrt{6}$.  Don't worry too much about drawing them to scale.  Just figure out what the side lengths need to be.

144. ★Draw an isosceles right triangle with a hypotenuse of $2\sqrt{2}$. Try to make your drawing look like an isosceles triangle and make sure it contains a right angle.

145. If you build a rectangular gate for a wooden fence, it will not be strong unless you use another wooden board to connect the opposite corners of the gate.

    a. If you build a gate that is 4 feet wide and 6 feet tall, how long must you cut a board if that board will connect the top right corner to the bottom left corner? Round your measurement to the nearest tenth of a foot.

    ★b. Round your measurement to the nearest tenth of an inch.

146. What is the value of x for the right triangle shown?

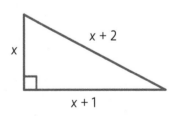

147. What is the perimeter of the triangle in the previous scenario?

148. A rectangular garden has a diagonal path that runs from one corner to the opposite corner. The width of the garden is 4 feet less than the length of the garden. If the length of the path is 20 feet, what are the dimensions of the garden?

149. A ladder leans against a wall. The distance from the top of the ladder to the ground is 2 more feet than the distance from the bottom of the ladder to the wall. The ladder is 10 feet long. Determine the distance from the top of the ladder to the ground.

150. You have now learned how to perform operations with irrational numbers, but you may have lost some of your familiarity with rational numbers. See if you remember how to add, subtract, multiply, and divide fractions as you simplify each expression below.

a. $\dfrac{2}{3}+\dfrac{2}{3}$

b. $\dfrac{8}{3}-\dfrac{4}{3}$

c. $\dfrac{6}{3}\cdot\dfrac{2}{3}$

d. $\dfrac{\frac{2}{3}}{2}$

151. Simplify the radical expressions below.

a. $\sqrt{2}+\sqrt{2}$

b. $5\sqrt{2}-3\sqrt{2}$

c. $\sqrt{4}\cdot\sqrt{2}$

d. $\dfrac{4}{\sqrt{2}}$

Use this page to record important ideas in the previous section or
for any other writing that helps you learn the topics in this book.

53

Section 11
# REVIEW OF RADICAL EXPRESSIONS

152. Multiply each group of expressions and simplify the result as much as you can. Write your answer in the form $A\sqrt{B}$, if possible.

a. $2\sqrt{3}\cdot 3\sqrt{3}$

b. $\left(-2\sqrt{5}\right)\left(\sqrt{5}\right)$

c. $\left(2\sqrt{2}\right)\left(-3\sqrt{3}\right)$

d. $\left(-4\sqrt{3}\right)^2$

153. Simplify the following expressions.

a. $\left(1+\sqrt{2}\right)\left(1-\sqrt{2}\right)$

b. $\left(2+\sqrt{3}\right)\left(5-\sqrt{3}\right)$

c. $\left(3+2\sqrt{5}\right)\left(3-2\sqrt{5}\right)$

154. Simplify each expression as much as possible.

a. $\sqrt{\dfrac{2}{8}}$

b. $\dfrac{\sqrt{16}}{\sqrt{3}}$

c. $\dfrac{5\sqrt{2}}{\sqrt{5}}$

155. Determine the value of $x$ that makes each equation true.

a. $\sqrt{x-4}=7$

b. $\sqrt{3x-11}=5$

156. ★Determine the value of $x$ that makes each equation true.

a. $\sqrt{x+7}=x+1$

b. $\sqrt{x-2}=x-2$

Section 12
# CUMULATIVE REVIEW

157. Consider the graph to the right.

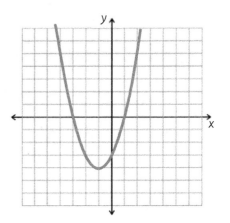

    a.  For which x-value(s) does the graph have a y-value of 0?

    b.  For which y-value(s) does the graph have an x-value of −1?

    c.  For which x-value(s) does the graph have a y-value of 5?

158. Determine the x-value(s) for which the equation $y=-x^2+4x+5$ has the y-value shown below?

    a.  0                                              b.  5

159. Identify any x- and y-intercepts for each of the following equations.

    a.  $2x-5y=-10$                              b.  $y=2x^2+5x-12$

160. The average (arithmetic mean) of 9, 10, 31, and R is 12. Determine the value of R.

161. The bookstore staff needed to order 90 sweatshirts, but when they checked the pricing for them, they found a special offer that gave a 15% discount on orders of 100 or more. To save money, they altered the intended order and purchased 100 sweatshirts. The order total came to $5100.

    a. What was the original price of a single sweatshirt?

    b. How much money did they save by purchasing 100 sweatshirts instead of 90?

162. Consider the ordered pairs (–5, –2) and (7, –6).

    a. Determine the equation of the line that passes through the given ordered pairs.

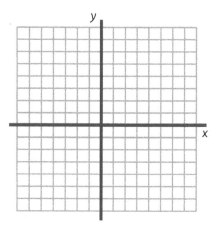

    b. Graph the line on the Cartesian plane shown.

163. While riding in the car on the highway one day, you look out the window and notice that the car is passing mile marker 82. You look at the clock and notice that it is 1:35pm. A little while later, after thinking about how slow the car seems to be moving, you look out the window again and notice mile marker 58. It is now 2:20pm. What was the average speed of the car over that time period, measured in miles per hour?

Use this page to record important ideas in the previous section or
for any other writing that helps you learn the topics in this book.

59

## Section 13
# ANSWER KEY

| | |
|---|---|
| 1. | a. 49   b. 125   c. 256   d. 32 |
| 2. | a. 7   b. 5   c. 4   d. 2 |
| 3. | a. square root   b. cube root<br>c. 4th root   d. 5th root   e. *n*th root |
| 4. | Row 1: no blanks need to be filled<br>Row 2: fill in both blanks with 125<br>Row 3: fill in both blanks with 256<br>Row 4: fill in both blanks with 32<br>Row 5: fill in both blanks with *B* |
| 5. | a. $\sqrt{81}$   b. $\sqrt[3]{64}$   c. $\sqrt[4]{81}$   d. $\sqrt[5]{1}$ |
| 6. | a. 8   b. 27   c. 64   d. 125 |
| 7. | a. 2   b. 3   c. 4   d. 5 |
| 8. | a. 16   b. 81   c. 256   d. 625 |
| 9. | a. 2   b. 3   c. 4   d. 5 |
| 10. | a. 9   b. 4   c. 3   d. 1 |
| 11. | a. 3   b. 2   c. 1   d. 0 |
| 12. | a. 4   b. −2   c. −4   d. not possible |
| 13. | You can check that your square root is accurate by squaring your result.  Based on what you have learned so far, it is not possible for a number to be negative <u>after</u> it has been squared. |
| 14. | a. $\dfrac{4}{9}$   b. $\dfrac{8}{27}$   c. $\dfrac{4}{9}$   d. $-\dfrac{8}{27}$ |
| 15. | a. $-\dfrac{3}{5}\cdot\dfrac{3}{5}\cdot\dfrac{3}{5}\rightarrow -\dfrac{27}{125}$   b. $\dfrac{x^8}{10{,}000}$<br><br>c. $\dfrac{36x^2}{121}$   d. $-\dfrac{32}{x^5 y^{15}}$ |
| 16. | $-\dfrac{2}{3}$ |
| 17. | a. $\dfrac{2}{3}$   b. $-\dfrac{1}{2}$   c. $-\dfrac{4}{5}$   d. not possible |
| 18. | You can check that your square root is accurate by squaring your result.  Based on what you have learned so far, it is not possible for a number to be negative <u>after</u> it has been squared. |
| 19. | $\sqrt{10}$ ; cannot simplify this more |
| 20. | 90 |

| | |
|---|---|
| 21. | a. $\dfrac{4}{20}\rightarrow\dfrac{1}{5}$   b. $\dfrac{6}{42}\rightarrow\dfrac{1}{7}$   c. $\dfrac{8}{72}\rightarrow\dfrac{1}{9}$ |
| 22. | a. ≈6.1   b. ≈7.7   c. ≈9.9 |
| 23. | a. 6.083   b. 7.746   c. 9.899 |
| 24. | No |
| 25. | Yes.  For example, $7=\dfrac{7}{1}$. |
| 26. | a. $\dfrac{1}{2}$   b. $\dfrac{1}{4}$   c. $\dfrac{1}{5}$   d. $\dfrac{3}{4}$ |
| 27. | a. $\dfrac{1}{3}$   b. $\dfrac{1}{9}$   c. $\dfrac{2}{3}$   d. $\dfrac{5}{9}$ |
| 28. | a. 3.14 → 3.1415926…<br>b. Calculators will vary → 6.08276253… |
| 29. | Circle $\sqrt{3}$ , $\sqrt{8}$ , and $\sqrt{15}$ . |
| 30. | 90 |
| 31. | a. 6   c. 10   e. 20   g. 12<br>b. 6   d. 10   f. 20   h. 12 |
| 32. | a. 8   c. 70   e. 48   g. 45<br>b. 8   d. 70   f. 48   h. 45 |
| 33. | a. $\sqrt{A}\cdot\sqrt{B}$   b. $\sqrt{X\cdot Y}$ |
| 34. | $5\sqrt{3}$ |
| 35. | $2\sqrt{5}$ |
| 36. | a. five root three   b. seven root six<br>c. four root ten |
| 37. | a. $2\sqrt{3}$   b. $3\sqrt{2}$   c. $5\sqrt{2}$   d. $\sqrt{10}$ |
| 38. | a. $3\sqrt{3}$   b. 3   c. two |
| 39. | a. $3\sqrt{2}$   b. $\sqrt{16}\sqrt{3}\rightarrow 4\sqrt{3}$<br>c. $\sqrt{4}\sqrt{5}\rightarrow 2\sqrt{5}$ |
| 40. | 1, 4, 9, 16, 25, 36, 49, 64, 81, 100 |
| 41. | a. 16   b. 9   c. 100   d. 64 |
| 42. | a. 4 and 16   b. 4 and 16<br>c. 4, 9 and 36   d. 4, 9, 16, 36 and 144 |
| 43. | a. $\sqrt{4}\sqrt{5}\rightarrow 2\sqrt{5}$   b. $\sqrt{4}\sqrt{6}\rightarrow 2\sqrt{6}$<br>c. $\sqrt{4}\sqrt{7}\rightarrow 2\sqrt{7}$ |
| 44. | a. $\sqrt{9}\sqrt{5}\rightarrow 3\sqrt{5}$   b. $\sqrt{9}\sqrt{6}\rightarrow 3\sqrt{6}$ |

# GUIDED DISCOVERY SCENARIOS

| # | Answer |
|---|--------|
| 45. | c. $\sqrt{9}\sqrt{7} \to 3\sqrt{7}$ <br> a. $\sqrt{25}\sqrt{2} \to 5\sqrt{2}$  b. $\sqrt{25}\sqrt{3} \to 5\sqrt{3}$ <br> c. $\sqrt{22}$ |
| 46. | a. $\sqrt{49}\sqrt{2} \to 7\sqrt{2}$  b. $\sqrt{4}\sqrt{10} \to 2\sqrt{10}$ |
| 47. | a. $\sqrt{81}\sqrt{2} \to 9\sqrt{2}$  b. $\sqrt{4}\sqrt{26} \to 2\sqrt{26}$ |
| 48. | 8 |
| 49. | 99 |
| 50. | a. $2\cdot2\sqrt{2} \to 4\sqrt{2}$  b. $3\cdot2\sqrt{3} \to 6\sqrt{3}$ |
| 51. | a. $7\cdot\sqrt{16}\sqrt{2} \to 7\cdot4\sqrt{2} \to 28\sqrt{2}$ <br> b. $5\cdot3=15$  c. $2\sqrt{14}$ |
| 52. | a. 1  b. 25  c. 8 |
| 53. | $H^2$ |
| 54. | $L^2$ |
| 55. | a. subtracting 3.1  b. multiplying by 3 <br> c. squaring an expression (raising an expression to the second power) |
| 56. | Raise 10 to the 2nd power (compute $10^2$) |
| 57. | $z-2=P^2$ and $z=P^2+2$ |
| 58. | a. 17  b. 0  c. 2 |
| 59. | no solution |
| 60. | a. 101  b. 6  c. 8  d. 24 |
| 61. | a. 27  b. −3  c. no solution |
| 62. | a. $(x+2)(x+2) \to x^2+4x+4$ <br> b. $(x-1)(x-1) \to x^2-2x+1$ <br> c. $x^2+10x+25$  d. $9x^2-42x+49$ |
| 63. | a. $(x+3)(x+2)=0 \to x=-3$ or $-2$ <br> b. $(x-6)(x+3)=0 \to x=6, -3$ <br> c. $(2x+3)(x-2)=0 \to x=-1.5, 2$ |
| 64. | a. $0=x^2-10x+16 \to 0=(x-8)(x-2)$ <br> $\to x=8$ ($x=8$ or 2, but $x=2$ makes the original equation false) <br> b. $0=x^2+9x+8 \to 0=(x+8)(x+1)$ <br> $\to x=-1$ ($x=-8$ or $-1$, but $x=-8$ makes the original equation false) |
| 65. | a. $\left(\sqrt{x+3}\right)^2=(x+1)^2 \to x+3=x^2+2x+1$ <br> $\to 0=x^2+x-2 \to 0=(x+2)(x-1)$ <br> $x=-2$ or 1 <br> b. $x=4$ or 3 |
| 66. | a. 1 is a solution, but −2 is **not** because it makes the original equation false <br> b. 4 and 3 are both solutions |
| 67. | $x=-1$ ($x=-9$ or $-1$, but $-9$ makes the |

| # | Answer |
|---|--------|
| | original equation false) |
| 68. | a. $x=8$ ($x=3$ or 8, but $x=3$ makes the original equation false) <br> b. no solution ($x=-1$, but that value makes the original equation false) |
| 69. | no solution ($x=1$ or 4, but both make the original equation false) |
| 70. | $N$ is 36 |
| 71. | a. $\sqrt{12} \to 2\sqrt{3}$  b. 5  c. 10 <br> d. $2\sqrt{2}\cdot\sqrt{14} \to 2\sqrt{28} \to 2\cdot2\sqrt{7} \to 4\sqrt{7}$ |
| 72. | a. 24  b. $6x$  c. $6\sqrt{2}$  d. $2A\sqrt{B}$ |
| 73. | a. $20\sqrt{3}$  b. 12  c. $6\sqrt{8} \to 12\sqrt{2}$ |
| 74. | a. $4\cdot3\sqrt{3} \to 12\sqrt{3}$  b. $2\sqrt{16} \to 8$ |
| 75. | a. 48  b. $15xy$  c. $15\sqrt{6}$ |
| 76. | a. $44\sqrt{14}$  b. $AC\sqrt{BD}$  c. $4\sqrt{4} \to 8$ |
| 77. | a. $40\sqrt{2}$  b. $-30\sqrt{3}$  c. −36  d. 63 |
| 78. | $\sqrt{18}=3\sqrt{2} \to 3(1.41)=4.23$ |
| 79. | $\sqrt{75}=5\sqrt{3} \to 5(1.73)=8.65$ |
| 80. | a. $x=7$  b. $x=23$  c. $x=-1$ |
| 81. | $G=24; H=48$ |
| 82. | a. $3\sqrt{2}, \sqrt{19}, 2\sqrt{5}$ <br> $\left(\sqrt{18}, \sqrt{19}, \sqrt{20}\right)$ <br> b. $3\sqrt{10}, 7\sqrt{2}, 6\sqrt{3}, 4\sqrt{7}$ <br> $\left(\sqrt{90}, \sqrt{98}, \sqrt{108}, \sqrt{112}\right)$ |
| 83. | a. $\sqrt{5}$  c. $2\sqrt{2}$  e. 4 <br> b. 3  d. 2  f. $2+2\sqrt{3}$ |
| 84. | No. $2 \neq \sqrt{2}$ |
| 85. | a. 5  c. $5\sqrt{3}$  e. 6 <br> b. 7  d. 5  f. 2 |
| 86. | a. No  b. No |
| 87. | a. $6xy$  b. $8MN$  c. $30x^3y^3$ |
| 88. | a. $\sqrt{4}=2$. Thus, $\sqrt{4}+\sqrt{4}=4$. <br> b. $\sqrt{4}=2; \sqrt{9}=3$. Thus, $\sqrt{4}+\sqrt{9}=5$. <br> c. $\sqrt{36}=6; \sqrt{25}=5$. Thus, $\sqrt{36}-\sqrt{25}=1$. |
| 89. | a. 5 eggs  b. $5x$  c. $5x^2$  d. $5\sqrt{2}$ |
| 90. | a. $2\sqrt{2}$  b. $2\sqrt{3}$  c. $3\sqrt{5}$ |
| 91. | a. $2\sqrt{2}$  b. $3\sqrt{3}$  c. $-4\sqrt{5}$ |

| | |
|---|---|
| 92. | $\sqrt{5}+\sqrt{6}$ cannot be simplified more. |
| 93. | "Like terms" have the same value underneath the radical symbol. |
| 94. | $\sqrt{20}+\sqrt{45}\rightarrow 2\sqrt{5}+3\sqrt{5}\rightarrow 5\sqrt{5}$ |
| 95. | $\sqrt{75}+\sqrt{48}\rightarrow 5\sqrt{3}-4\sqrt{3}\rightarrow 1\sqrt{3}=\sqrt{3}$ |
| 96. | a. $2\sqrt{2}-2\sqrt{2}\rightarrow 0$  b. $-\sqrt{5}+4\sqrt{5}=3\sqrt{5}$<br>c. $2\sqrt{3}-2\sqrt{3}\rightarrow 0$ |
| 97. | a. $\sqrt{2}+2\sqrt{2}\rightarrow 3\sqrt{2}$<br>b. $-\sqrt{10}+2\sqrt{10}\rightarrow 1\sqrt{10}$ or $\sqrt{10}$ |
| 98. | a. $2\sqrt{5}$     b. $-\sqrt{6}$     c. $3\sqrt{2}+3\sqrt{3}$ |
| 99. | a. $0$     b. $9\sqrt{3}-2\sqrt{5}$ |
| 100. | a. $3x+15$     b. $w^3-11w$<br>c. $-2y^5+14y^4-2y^3$ |
| 101. | a. $6+3\sqrt{3}$   b. $\sqrt{12}-4\sqrt{6}\rightarrow 2\sqrt{3}-4\sqrt{6}$<br>c. $\sqrt{5}-\sqrt{50}\rightarrow \sqrt{5}-5\sqrt{2}$ |
| 102. | a. $x^2+x-6$     b. $x^2-4x+4$<br>c. $4x^2-25$     d. $9x^2+6x+1$ |
| 103. | a. $3+2\sqrt{2}$   b. $19-8\sqrt{3}$   c. $29+12\sqrt{5}$ |
| 104. | a. $2$     b. $-4$     c. $1$ |
| 105. | After multiplying the binomials, the irrational terms are opposites and combine to make 0. They "cancel out." |
| 106. | a. $-1$     b. $4-2\sqrt{3}$     c. $-11$<br>a and c are rational |
| 107. | a. $27+10\sqrt{2}$     b. $-19$     c. $12+6\sqrt{3}$<br>b is rational |
| 108. | a. $\sqrt{2}$   b. $\sqrt{3}$   c. $5+\sqrt{3}$   d. $1-3\sqrt{2}$ |
| 109. | No. Examples of perfect square fractions include $\frac{1}{4}, \frac{1}{9}, \frac{4}{9}, \frac{16}{25}$, etc... |
| 110. | $4\left(3\sqrt{2}\right)$ or $12\sqrt{2}$ inches |
| 111. | a. $\frac{1}{2}$   b. $\frac{2}{3}$   c. $\frac{5}{3}$   d. $\frac{1}{2}$ |
| 112. | a. $1$   b. $1$   c. $2$   d. $6$ |
| 113. | a. $\frac{4}{2}=2$     b. $\frac{9}{3}=3$<br>c. $\sqrt{4}=2$     d. $\sqrt{9}=3$ |
| 114. | a. $\frac{2\sqrt{5}}{2}=\sqrt{5}$     b. $\frac{3\sqrt{6}}{3}=\sqrt{6}$<br>c. $\sqrt{5}$     d. $\sqrt{6}$ |

| | |
|---|---|
| 115. | a. $\frac{\sqrt{A}}{\sqrt{B}}$     b. $\sqrt{\frac{X}{Y}}$ |
| 116. | a. $\frac{\sqrt{5}}{2}$     b. $\frac{\sqrt{2}}{3}$     c. $\frac{8\sqrt{3}}{5}$ |
| 117. | a. $\frac{1}{2\sqrt{2}}$     b. $\frac{3}{\sqrt{5}}$     c. $\frac{12\sqrt{2}}{\sqrt{3}}$ |
| 118. | There is a radical expression in each denominator in the previous scenario, but **not** in scenario before that. |
| 119. | $3$     $y$     $x+3$ |
| 120. | $1$ |
| 121. | $\frac{1}{2}+\frac{1}{6}=\frac{3}{6}+\frac{1}{6}=\frac{4}{6}$ or $\frac{2}{3}$ |
| 122. | $\frac{2x}{x}-\frac{3}{x}=\frac{2x-3}{x}$ |
| 123. | $\frac{3\sqrt{2}}{2}+\frac{1}{\sqrt{2}}=\frac{3\sqrt{2}}{2}+\frac{\sqrt{2}}{2}=\frac{4\sqrt{2}}{2}$ or $2\sqrt{2}$ |
| 124. | a. $\frac{\sqrt{2}}{\sqrt{2}}$   b. $\frac{\sqrt{6}}{\sqrt{6}}$   c. $\frac{\sqrt{3}}{\sqrt{3}}$   d. $\frac{\sqrt{5}}{\sqrt{5}}$ |
| 125. | a. $\frac{4\sqrt{3}}{3}$   b. $\sqrt{5}$   c. $2\sqrt{7}$   d. $4\sqrt{6}$ |
| 126. | You will still have $\frac{2\sqrt{3}}{9}$ either way, so it is easier to multiply by $\frac{\sqrt{3}}{\sqrt{3}}$. |
| 127. | a. $\frac{6}{\sqrt{11}}\cdot\frac{\sqrt{11}}{\sqrt{11}}\rightarrow\frac{6\sqrt{11}}{11}$<br>b. $\frac{10}{\sqrt{2}}\cdot\frac{\sqrt{2}}{\sqrt{2}}\rightarrow\frac{10\sqrt{2}}{2}\rightarrow 5\sqrt{2}$<br>c. $\frac{6}{5\sqrt{2}}\cdot\frac{\sqrt{2}}{\sqrt{2}}\rightarrow\frac{6\sqrt{2}}{10}\rightarrow\frac{3\sqrt{2}}{5}$ |
| 128. | a. $\frac{4\sqrt{3}}{3}$     b. $\frac{\sqrt{2}}{2}$     c. $\frac{3\sqrt{5}}{10}$ |
| 129. | a. $\frac{4\sqrt{3}}{10}+\frac{15\sqrt{3}}{10}\rightarrow\frac{19\sqrt{3}}{10}$<br>b. $\frac{2\sqrt{3}+5\sqrt{3}}{3}\rightarrow\frac{7\sqrt{3}}{3}$ |
| 130. | a. $\frac{3\sqrt{2}+2\sqrt{3}}{6}$     b. $\frac{6\sqrt{5}+5\sqrt{3}}{15}$ |
| 131. | a. 5 cm     b. 12 cm, 13 cm |
| 132. | $a^2+b^2=c^2$ |
| 133. | a. $10$   b. $\sqrt{10}$   c. $2\sqrt{10}$   d. $6$ |

| | |
|---|---|
| 134. | a. $x = 2\sqrt{5}$    b. $x = 13$    c. $x = 2$ |
| 135. | The commutative property of addition:<br>$a + b = b + a$ |
| 136. | a. A and C    b. D and E    c. E and G |
| 137. | hypotenuse |
| 138. | a. $2\sqrt{10}$    b. $2\sqrt{5}$    c. $5\sqrt{2}$ |
| 139. | $\sqrt{a^2 + b^2}$ |
| 140. | a. $2\sqrt{5}$ (4.5cm)    b. $2\sqrt{13}$ (7.2cm) |
| 141. | a. $\sqrt{3}$    b. 10    c. 2 |
| 142. | $M = 5\sqrt{3}$ or $\approx 8.7$ ; solve $5^2 + M^2 = 10^2$ |
| 143. | The squares of the side lengths must add up to 6 |
| 144. | The squares of the side lengths must add up to 8. The triangle should look like a square cut in half along its diagonal. |
| 145. | a. 7.2 feet    b. 7 feet, 2.5 inches |
| 146. | $x = 3$; solve: $x^2 + (x+1)^2 = (x+2)^2$<br>$x^2 + x^2 + 2x + 1 = x^2 + 4x + 4 \to x^2 - 2x - 3 = 0$<br>$(x-3)(x+1) = 0 \to x = 3$ or $-1$ |
| 147. | 3 + 4 + 5 = 12 units |
| 148. | width: 12 feet; length: 16 feet<br>solve: $L^2 + (L-4)^2 = 20^2$<br>or $w^2 + (w+4)^2 = 20^2$ |
| 149. | 8 feet; solve $x^2 + (x+2)^2 = 10^2$ |

| | |
|---|---|
| | $x^2 + x^2 + 4x + 4 = 100 \to 2x^2 + 4x - 96 = 0$<br>$x^2 + 2x - 48 = 0 \to (x-6)(x+8) = 0$<br>$x = 6$ or $-8$ |
| 150. | Each expression simplifies to $\dfrac{4}{3}$. |
| 151. | Each expression simplifies to $2\sqrt{2}$. |
| 152. | a. 18    b. $-10$    c. $-6\sqrt{6}$    d. 48 |
| 153. | a. $-1$    b. $7 + 3\sqrt{3}$    c. $-11$ |
| 154. | a. $\dfrac{1}{2}$    b. $\dfrac{4\sqrt{3}}{3}$    c. $\sqrt{10}$ |
| 155. | a. $x - 4 = 49 \to x = 53$<br>b. $3x - 11 = 25 \to 3x = 36 \to x = 12$ |
| 156. | a. $x = -3$ or 2 ($-3$ makes the original equation false)    b. $x = 2$ or 3 |
| 157. | a. $-3, 1$    b. $-4$    c. $-4, 2$ |
| 158. | a. $-1, 5$    b. 0, 4 |
| 159. | a. x-int: $(-5,0)$; y-int: $(0,2)$<br>b. x-int: $(1.5,0)$, $(-4,0)$; y-int: $(0,-12)$ |
| 160. | $-2$ |
| 161. | a. \$60    b. \$300 (\$5100 for 100 instead of \$5400 for 90) |
| 162. | $y = -\dfrac{1}{3}x - \dfrac{11}{3}$ |
| 163. | 24 miles in 0.75 hours = 32 mph |

64

# HOMEWORK & EXTRA PRACTICE SCENARIOS

As you complete scenarios in this part of the book, you will practice what you learned in the guided discovery sections. You will develop a greater proficiency with the vocabulary, symbols and concepts presented in this book. Practice will improve your ability to retain these ideas and skills over longer periods of time.

There is an Answer Key at the end of this part of the book. Check the Answer Key after every scenario to ensure that you are accurately practicing what you have learned. If you struggle to complete any scenarios, try to find someone who can guide you through them.

# CONTENTS

# Section 1
# REVIEW

1. Perform the operation shown and simplify the result as much as you can.

   a. $\dfrac{2}{3} \cdot \dfrac{9}{4}$         b. $\dfrac{2}{3} + \dfrac{9}{4}$         c. $\dfrac{2}{3} \div \dfrac{9}{4}$         d. $\dfrac{2}{3} - \dfrac{9}{4}$

2. What is the Slope-Intercept Form for the equation of a line?

3. What is the Standard Form for the equation of a line?

4. Write the equation of the line shown, in Slope-Intercept Form.

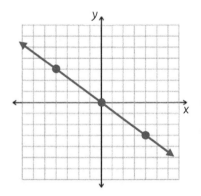

5. In the same *xy*-plane as the previous scenario, graph the line formed by the equation $4x - 3y = -15$.

6. Find the exact location of the intersection point of the previous two lines.

## Section 2
# EXPONENTS AND ROOTS

69

7. A root is the <u>inverse</u> operation for an exponent because the purpose of a root is to <u>undo</u> the result of an exponent. In order to accurately compute roots, you need to be able to think about exponents. Fill in the following table. If you need help, you may use a calculator.

| | | |
|---|---|---|
| $2^2 =$ | $2^3 =$ | $2^4 =$ |
| $(-3)^2 =$ | $(-3)^3 =$ | $(-3)^4 =$ |
| $4^2 =$ | $4^3 =$ | $4^4 =$ |

8. Fill in the following table. If you need help, you may use a calculator.

| | | |
|---|---|---|
| $(-5)^2 =$ | $(-5)^3 =$ | $(-5)^4 =$ |
| $6^2 =$ | $6^3 =$ | $6^4 =$ |
| $(-7)^2 =$ | $(-7)^3 =$ | $(-7)^4 =$ |

9. Simplify the following expressions by raising each base to its specified exponent.

a. $\left(-2\right)^5$  b. $9^2$  c. $\left(-8\right)^3$

In the previous scenario, you do a specific operation: raise a number to a specific exponent by multiplying that number by itself a specific number of times. If there exists an inverse operation for this, then it must <u>undo</u> the work of that multiplication.

10. Answer the following questions.

    a. What number when squared (raised to the 2nd power) is equal to 49?

    b. What number when cubed (raised to the 3rd power) is equal to 125?

    c. What number when raised to the 4th power is equal to 256?

    d. What number when raised to the 5th power is equal to 32?

11. Use what you have learned so far to answer each of the following questions.

    a. What is the square root of 121?

    b. What is the cube root of 27?

    c. What is the fourth root of 16?

12. Think about multiples of 0 and 1 as you consider each question below.

    a. What is the fifth root of 1?        b. What is the fifth root of 0?

13. Answer each of the questions below.

    a. What is the cube root of −216?        b. What is the fifth root of −100,000?

14. Answer each of the questions below.

    a. What is the square root of −36?        b. What is the fourth root of −256?

15. In the previous scenario, why is not possible to find the square root or the fourth root of a negative number?

16. Compute the following roots. Do not spend much time on each one. If it helps, try to find the numbers in the work you did in an earlier scenario.

   a. $\sqrt{1}$    b. $\sqrt[3]{-27}$    c. $\sqrt[4]{625}$    d. $\sqrt[5]{-32}$

17. Compute the following roots. Do not spend much time on each one. If it helps, try to find the numbers in the work you did in an earlier scenario.

   a. $\sqrt[5]{-100,000}$    b. $\sqrt[4]{1296}$    c. $\sqrt[3]{-343}$    d. $\sqrt{-25}$

18. In the previous scenario, explain why the value of $\sqrt{-25}$ is not $-5$?

19. Apply what you know about fractions to simplify each expression.

   a. $\left(\dfrac{2}{3}\right)^3$    b. $\left(-\dfrac{1}{7}\right)^2$    c. $\left(-\dfrac{5}{2}\right)^3$

20. Simplify the following expressions (apply the exponent to remove the parentheses).

   a. $\left(\dfrac{x}{2}\right)^4$    b. $\left(-\dfrac{3y}{4}\right)^2$    ★c. $\left(-\dfrac{2}{x^2y}\right)^5$

21. The expression $\left(-\dfrac{5}{2}\right)^3$ is equal to $-\dfrac{125}{8}$. It follows that the cube root of $-\dfrac{125}{8}$, written as $\sqrt[3]{-\dfrac{125}{8}}$, must be $-\dfrac{5}{2}$. What is the simplified form of $\sqrt[4]{\dfrac{1}{16}}$ ?

22. Compute the following roots.

    a. $\sqrt[4]{\dfrac{1}{625}}$
    b. $\sqrt[3]{-\dfrac{27}{64}}$
    c. $\sqrt[5]{-\dfrac{243}{100,000}}$
    d. $\sqrt{-\dfrac{16}{25}}$

23. In the previous scenario, explain why you cannot compute the value of $\sqrt{-\dfrac{16}{25}}$?

24. Simplify each root shown.

    a. $\sqrt{\dfrac{9y^2}{16}}$
    b. $\sqrt[4]{\dfrac{x^4}{16}}$
    c. $\sqrt[3]{-\dfrac{27x^3}{1,000y^6}}$

25. Without using a calculator, what is the cube root of 2?

26. Write the <u>perfect cubes</u> that are greater than 0 and less than 500.  How many are there?

27. How many integers from 1 to 1000, inclusive, are <u>not</u> perfect cubes?

28. What is the probability that an integer will be a perfect <u>cube</u> if you randomly select an integer from the following range?

    a. 1 to 8
    b. 1 to 125
    ★c. 1 to 1,000

29. Without a calculator, estimate the value of each cube root, rounded to the nearest tenth.

     a. $\sqrt[3]{9}$              b. $\sqrt[3]{70}$              c. $\sqrt[3]{120}$

30. Use a calculator to determine the following cube roots with a calculator. Round your answers to two decimal places.

     a. $\sqrt[3]{9}$              b. $\sqrt[3]{70}$              c. $\sqrt[3]{120}$

31. Consider the first 1000 cube roots of integers $\left(\sqrt[3]{1}, \sqrt[3]{2}, \sqrt[3]{3}, \ldots, \sqrt[3]{1000}\right)$. How many are irrational?

32. Write the perfect squares that are greater than 0 and less than 100. How many are there?

33. Without a calculator, estimate the value of each square root, rounded to the nearest tenth.

     a. $\sqrt{11}$              b. $\sqrt{77}$              c. $\sqrt{50}$

34. Four radical expressions are listed below. Mark their approximate location on the number line.

     $\sqrt{20}, \sqrt{40}, \sqrt{60}, \sqrt{80}$

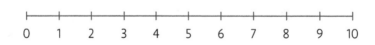

*Section 3*

# SIMPLIFYING RADICAL EXPRESSIONS

The next scenarios use your familiarity with rational square roots to help you learn how to simplify square roots that are irrational.

35. Simplify each product as much as possible. Analyze the results as you work through each column.

a. $\sqrt{25 \cdot 36}$

c. $\sqrt{49 \cdot 4}$

e. $\sqrt{16 \cdot 49}$

b. $\sqrt{25} \cdot \sqrt{36}$

d. $\sqrt{49} \cdot \sqrt{4}$

f. $\sqrt{16} \cdot \sqrt{49}$

36. Complete the following statements.

a. $\sqrt{A \cdot B}$ is equivalent to ...

b. $\sqrt{X} \cdot \sqrt{Y}$ is equivalent to ....

37. What is the most simplified form of $\sqrt{100} \cdot \sqrt{3}$ ?

38. The expression $2\sqrt{5}$ means "2 multiplied by the square root of 5." Another way to say this is "2 times the square root of 5." In order to make it easier to say, you can shorten this to say "two root five."

The expression $4\sqrt{11}$ can be written out with words as "four root eleven." Write out each of the following expressions using only words.

a. $4\sqrt{2}$

b. $2\sqrt{5}$

c. $9\sqrt{7}$

39. Simplify each product as much as possible.

a. $\sqrt{16} \cdot \sqrt{2}$

b. $\sqrt{5} \cdot \sqrt{4}$

40. Simplify each product as much as possible.

a. $\sqrt{7} \cdot \sqrt{81}$

b. $\sqrt{3} \cdot \sqrt{11}$

41. Write down the first twelve perfect squares.

42. What is the most simplified form of $\sqrt{12}$ ?

43. Radical expressions are simplified below. Fill in the missing values in each scenario.

a. $\sqrt{54}$
$\sqrt{9 \cdot 6}$
$\sqrt{9}\sqrt{6}$
$\underline{\quad}\sqrt{6}$

b. $\sqrt{80}$
$\sqrt{\underline{\quad} \cdot 5}$
$\sqrt{\underline{\quad}}\sqrt{5}$
$\underline{\quad}\sqrt{5}$

c. $\sqrt{52}$
$\sqrt{\underline{\quad} \cdot \underline{\quad}}$
$\sqrt{\underline{\quad}}\sqrt{\underline{\quad}}$
$\underline{\quad}\sqrt{\underline{\quad}}$

44. Simplify the following expressions. Write your answer in the form $A\sqrt{B}$.

a. $\sqrt{24}$

b. $\sqrt{40}$

c. $\sqrt{30}$

45. What is the largest number that is a factor of the number shown and also a perfect square?

a. 50

b. 128

46. What is the largest perfect square factor of the number shown?

a. 200

b. 108

47. Simplify each expression. Write your answer in the form $A\sqrt{B}$.

a. $\sqrt{28}$

b. $\sqrt{32}$

48. Simplify each expression. Write your answer in the form $A\sqrt{B}$.

    a. $\sqrt{60}$                                    b. $\sqrt{29}$

49. Fill in the blank. The square root of _____ is $3\sqrt{3}$.

50. If the square root of a number is $B$, what is the number?

51. The square root of a number is shown below. What is the original number?

    a. $3\sqrt{5}$             b. $2\sqrt{22}$             c. $5\sqrt{10}$

52. Some expressions may look simplified, but they contain expressions that still contain perfect square factors. Simplify each of the following expressions completely.

    a. $9\sqrt{4}$             b. $3\sqrt{20}$             c. $2\sqrt{27}$

53. Simplify each of the following expressions as much as you can.

    a. $7\sqrt{8}$             b. $5\sqrt{72}$             c. $7\sqrt{22}$

*Section 4*
# SOLVING EQUATIONS WITH RADICALS

54. Determine the value of $y$ that makes each equation true.

    a.  $\sqrt{y}=2$        b.  $\sqrt{y}=7$        c.  $\sqrt{y}=5\sqrt{2}$        d.  $\sqrt{y}=-5$

55. If $\sqrt{y}=M$, then $y =$ _____.

56. If $\sqrt{x+5}=T$, then $x =$ _____.

57. Determine the value of $x$ that makes each equation true.

    a.  $\sqrt{x-3}=4$        b.  $\sqrt{5x-1}=3$        c.  $\sqrt{x}=-7$

58. If $\sqrt{x}=-3$, then $x =$ _____.

In the previous scenario, the equation has no solution, but this may be confusing. If you replace $x$ with 25, the equation becomes $\sqrt{9}=-3$, or 3 = −3, which is a false statement.

59. Solve each equation. Notice how the structure of these equations is slightly different than the equations in the previous scenario.

    a.  $1+\sqrt{x+5}=3$        b.  $5+\sqrt{2-x}=6$       ★c.  $7-\sqrt{1-2x}=2$

60. Determine the value of $x$ that makes each equation true.

    a.  $\sqrt{x+8}=\sqrt{11}$      ★b.  $\sqrt{x-10}=2\sqrt{2}$      ★c.  $\sqrt{7x-4}=2\sqrt{6}$

61. Simplify each expression.

    a. $(x-8)^2$                                     b. $(5-2x)^2$

62. Solve the equations.

    a. $x^2-7x+12=0$                          ★b. $4x^2-17x-15=0$

63. The following equation is partially solved for you. Fill in the blank shown and then keep going until you have solved the equation.

$$\sqrt{4x^2-3x}=2x-1 \rightarrow \left(\sqrt{4x^2-3x}\right)^2=(2x-1)^2$$
$$4x^2-3x=\underline{\hspace{3cm}}$$

64. Use what you just reviewed to solve each equation.

    a. $\sqrt{2x}=x-4$                            ★b. $\sqrt{5+x}=x-7$

When you square both sides in the previous scenario, the right side of the equation becomes a trinomial. After you bring all of the terms to one side of the equation, you can factor the expression. Your two separate factors result in two separate solutions (unless the factors are identical).

65. You may have assumed that both solutions made the original equation true. If you have not already done this, check your solutions by plugging them back into the original equation. Confirm that each solution makes the original equation a true statement.

66. Solve the equation $\sqrt{x^2+40x}=x+10$.

67. When a number, $N$, is decreased by 7, the square root of the result is 3. What is the value of $N$?

68. After a number is increased by 1, the square root of the result is 0.5. What is the number?

69. ★After the area of a square is increased by 5 cm$^2$, the side lengths are now 3 cm. What was the original perimeter of the square? Drawing a square or two may help you organize your work.

70. ★When the value of $K$ decreases by 5, the cube root of the result is –2. What is $K$?

71. If $\sqrt{K}=2\sqrt{5}$, $\sqrt{L}=3\sqrt{5}$, $\sqrt{M}=4\sqrt{5}$ what is the value of $K+L+M$?

72. Without a calculator, find a way to arrange the following numbers from smallest to largest.

$2\sqrt{21}$, $3\sqrt{9}$, $4\sqrt{5}$, $5\sqrt{3}$

73. Write out a set of directions that explain how to write the expression $\sqrt{54}$ in simplified form.

Section 5
# MULTIPLICATION WITH RADICALS

74. Apply what you have learned so far to simplify the following expressions.

 a. $5\sqrt{32}$

 b. $\sqrt{3}\cdot\sqrt{15}$

75. Apply what you have learned so far to simplify the following expressions.

 a. $\sqrt{2}\cdot\sqrt{2}$

 b. $\sqrt{30}\cdot\sqrt{12}$

 c. $\left(\sqrt{6}\right)^2$

76. Simplify each expression.

 a. $4(5y)$

 b. $x\left(x\sqrt{y}\right)$

 c. $\left(\sqrt{x}\right)^2$

77. Simplify each expression. Write your answer in the form $A\sqrt{B}$, if possible.

 a. $3\left(2\sqrt{2}\right)$

 b. $9\left(2\sqrt{7}\right)$

 ★c. $\sqrt{3}\left(6\sqrt{3}\right)$

78. Multiply each group of expressions and simplify the result as much as you can.

 a. $4x\cdot4y$

 b. $2\sqrt{7}\cdot4\sqrt{2}$

 c. $\left(2\sqrt{5}\right)^2$

 d. $x\sqrt{y}\cdot x\sqrt{y}$

79. Multiply each group of expressions. Write your answer in the form $A\sqrt{B}$, if possible.

    a. $\sqrt{50} \cdot 3\sqrt{6}$        b. $\left(-3\sqrt{8}\right)\left(5\sqrt{2}\right)$       c. $\left(3\sqrt{3}\right)\left(-\sqrt{6}\right)$       d. $\left(-2\sqrt{6}\right)^2$

80. ★Do <u>not</u> use a calculator. If $\sqrt{12}$ is 3.46 when rounded to 2 decimal places, what is the approximate value of $\sqrt{48}$ when rounded to 2 decimal places.

81. Solve each equation.

    a. $\dfrac{2}{3}y = y + 4$             b. $\sqrt{f-7} = 2$           c. $10 - \sqrt{x-1} = 4$

At this point, you have learned how to solve equations that contain radical expressions. You have also discovered properties that allow you to simplify radicals and multiply radicals. The next set of properties to consider are those that involve addition and subtraction.

*Section 6*
# ADDITION AND SUBTRACTION WITH RADICALS

82. Simplify each expression as much as possible.

   a. $\sqrt{100+25}$

   b. $\sqrt{100}+\sqrt{25}$

   c. $\sqrt{9-4}$

   d. $\sqrt{9}-\sqrt{4}$

   e. $\sqrt{12+24}$

   f. $\sqrt{12}+\sqrt{24}$

83. Is the expression $\sqrt{4}+\sqrt{4}$ equal to the expression $\sqrt{4+4}$ ?

84. Determine if the following properties exist.

   a. $\sqrt{x}+\sqrt{y}=\sqrt{x+y}$

   b. $\sqrt{x}-\sqrt{y}=\sqrt{x-y}$ .

85. Simplify each expression as much as you can.

   a. $1\sqrt{5}+1\sqrt{5}$

   b. $\sqrt{7}+\sqrt{7}$

   c. $\sqrt{11}+3\sqrt{11}$

86. Simplify each expression as much as you can.

   a. $\sqrt{3}+2\sqrt{3}$

   b. $\sqrt{10}-\sqrt{5}$

   c. $5\sqrt{2}+3\sqrt{2}$

   d. $\sqrt{7}-2\sqrt{14}$

87. Explain how you can determine if two radical expressions are like terms.

88. Simplify each expression as much as possible.

   a. $5\sqrt{6}-3\sqrt{6}$

   b. $-2\sqrt{5}+3\sqrt{5}$

89. Simplify the expression $\sqrt{99} + \sqrt{44}$.

90. Simplify each expression as much as possible.

    a. $-2\sqrt{7} + \sqrt{28}$

    b. $\sqrt{8} - \sqrt{12}$

91. Simplify each expression as much as possible.

    a. $\sqrt{12} + \sqrt{3}$

    b. $\sqrt{32} - \sqrt{8}$

92. Simplify each expression.

    a. $\sqrt{20} + \sqrt{24}$

    b. $\sqrt{63} - \sqrt{28}$

93. A rectangle has a length of $\sqrt{90}$ cm and a width of $\sqrt{160}$ cm.

    a. Calculate the perimeter of the rectangle.

    b. What is the area of the rectangle?

94. Simplify the expression $\sqrt{48} - \sqrt{108} + \sqrt{12}$ as much as possible.

*Section 7*

# THE DISTRIBUTIVE PROPERTY WITH RADICALS

95. Use the Distributive Property to simplify the following expressions.

a. $5(w-6)$

b. $-x(2x^2-1)$

c. $-3y^2(y^3+8y^2-10)$

96. Apply the Distributive Property to simplify the following expressions.

a. $2(3-\sqrt{5})$

b. $\sqrt{6}(2-\sqrt{3})$

c. $\sqrt{3}(\sqrt{12}+\sqrt{27})$

97. Simplify the following expressions.

a. $(2x+7)(x-2)$

b. $(2x-9)^2$

c. $(x+3y)(x-3y)$

98. Simplify the following expressions.

a. $(4-\sqrt{2})(4+\sqrt{2})$

b. $(-3+\sqrt{6})(-3-\sqrt{6})$

c. $(1+2\sqrt{2})(1-2\sqrt{2})$

99. Simplify the following expressions.

a. $(4-\sqrt{2})^2$

b. $(-3+\sqrt{6})^2$

c. $(1+2\sqrt{2})^2$

100. As a matter of review, if an expression is *completely* simplified and it *still* contains a radical expression, such as $\sqrt{10}$ or $3-\sqrt{2}$, the expression is identified as irrational. In the previous two scenarios, which expressions are rational? Explain why this happens.

101. Which of the following expressions are rational? If it is helpful, simplify each expression.

    a. $\left(4-\sqrt{11}\right)\left(4+\sqrt{11}\right)$      b. $\left(6+\sqrt{2}\right)^2$      c. $\left(-7-\sqrt{3}\right)\left(-7+\sqrt{3}\right)$

102. Which of the following expressions are rational? If it is helpful, simplify each expression.

    a. $\left(\sqrt{4}-1\right)^2$      b. $\left(5-3\sqrt{10}\right)\left(5+3\sqrt{10}\right)$      c. $\left(2-2\sqrt{6}\right)\left(1+6\sqrt{2}\right)$

103. Fill in the blank. Identify an expression that will make each product rational.

    a. $\sqrt{3}\cdot$____      b. $\sqrt{5}\cdot$____      c. $\left(7-\sqrt{2}\right)($_____$)$      d. $\left(3+5\sqrt{7}\right)($_____$)$

104. A square has an area of 8 in². How long is one of the sides of this square?

*Section 8*
# SIMPLIFYING RADICALS WITH FRACTIONS

105. Use what you have learned so far to simplify each root as much as possible.

a. $\sqrt{\dfrac{1}{36}}$
b. $\sqrt{\dfrac{16}{49}}$
c. $\sqrt{\dfrac{81}{4}}$
d. $\sqrt{\dfrac{121}{144}}$

106. Simplify each product as much as possible.

a. $\sqrt{\dfrac{7}{10}}\cdot\sqrt{\dfrac{10}{7}}$
b. $\sqrt{\dfrac{3}{4}}\cdot\sqrt{4}$
c. $\sqrt{5}\cdot\sqrt{\dfrac{9}{5}}$
d. $\sqrt{\dfrac{8}{3}}\cdot\sqrt{\dfrac{27}{2}}$

107. Working with fractions leads naturally to the concept of division. Find out if any patterns emerge when division is combined with square roots by simplifying each expression as much as possible and paying close attention to your results.

a. $\sqrt{\dfrac{100}{25}}$
b. $\sqrt{\dfrac{45}{9}}$

c. $\dfrac{\sqrt{100}}{\sqrt{25}}$
d. $\dfrac{\sqrt{45}}{\sqrt{9}}$

108. Simplify each expression and analyze the results.

a. $\sqrt{\dfrac{81}{16}}$
b. $\sqrt{\dfrac{160}{80}}$

c. $\dfrac{\sqrt{81}}{\sqrt{16}}$
d. $\dfrac{\sqrt{160}}{\sqrt{80}}$

109. Your work in the previous scenario should suggest the following property:

a. $\sqrt{\dfrac{x}{y}}$ is equivalent to ☐
b. $\dfrac{\sqrt{f}}{\sqrt{g}}$ is equivalent to ☐

110. Simplify each expression.

    a. $\dfrac{\sqrt{9}}{\sqrt{144}}$
        b. $\dfrac{\sqrt{121}}{\sqrt{16}}$
        c. $\dfrac{\sqrt{100}}{\sqrt{36}}$

111. Simplify each expression.

    a. $\sqrt{\dfrac{2}{18}}$
        b. $\sqrt{\dfrac{64}{16}}$
        c. $\sqrt{\dfrac{45}{9}}$

112. Simmplify each expression as much as possible.

    a. $\sqrt{\dfrac{3}{16}}$
        b. $\dfrac{\sqrt{6}}{\sqrt{25}}$
        c. $\dfrac{2\sqrt{18}}{\sqrt{32}}$

113. This may seem repetitive, but once again, simplify each scenario as much as possible.

    a. $\dfrac{\sqrt{4}}{\sqrt{3}}$
        b. $\sqrt{\dfrac{1}{12}}$
        c. $\dfrac{\sqrt{50}}{\sqrt{7}}$

114. After you have checked your results, look carefully at your simplified fractions in the previous two scenarios. Can you identify a structural difference between the fractions in those two scenarios?

Section 9
# HOW TO RATIONALIZE A DENOMINATOR

115. If you were to multiply each expression below by a form of 1 to make the denominator a rational number, what form of 1 would you use?  Each scenario will require a different form of 1.

    a. $\dfrac{1}{\sqrt{3}}$
    b. $\dfrac{2}{\sqrt{5}}$
    c. $\dfrac{3}{\sqrt{4}}$
    d. $\dfrac{4}{\sqrt{9}}$

116. Multiply each expression by a form of 1 that makes the denominator a rational number.  Simplify the result as much as you can.

    a. $\dfrac{2}{\sqrt{2}}$
    b. $\dfrac{\sqrt{3}}{\sqrt{4}}$
    c. $\dfrac{6}{2\sqrt{3}}$
    ★d. $\dfrac{15\sqrt{2}}{\sqrt{5}}$

117. Consider the expression $\dfrac{5}{\sqrt{8}}$ .

    a.  Multiply the expression by $\dfrac{\sqrt{8}}{\sqrt{8}}$ and simplify.

    b.  Now multiply the original expression by $\dfrac{\sqrt{2}}{\sqrt{2}}$ and simplify again.

    c.  What do you notice?

118. Simplify each expression.  Make the denominator rational in your final result.

    a. $\sqrt{\dfrac{1}{3}}$
    b. $\sqrt{\dfrac{9}{2}}$
    ★c. $\sqrt{\dfrac{5}{27}}$

119. Add the following fractions. Rationalize the denominator in your final answer.

a. $\dfrac{2\sqrt{5}}{6} + \dfrac{2\sqrt{5}}{3}$

b. $\dfrac{\sqrt{2}}{2} + \dfrac{1}{\sqrt{2}}$

c. $\dfrac{3}{\sqrt{3}} + \dfrac{\sqrt{3}}{5}$

120. ★Simplify the expression shown. Rationalize the denominator in your final answer.

$$\sqrt{\dfrac{1}{2}} + \sqrt{\dfrac{3}{2}}$$

121. You have now learned how to perform operations with irrational numbers, but you may have lost some of your familiarity with rational numbers. See if you remember how to add, subtract, multiply, and divide fractions as you simplify each expression below.

a. $\dfrac{3}{7} + \dfrac{3}{7}$

b. $\dfrac{11}{7} - \dfrac{5}{7}$

c. $\dfrac{21}{14} \cdot \dfrac{4}{7}$

d. $\dfrac{\frac{3}{7}}{2}$

122. Simplify the radical expressions below.

a. $2\sqrt{3} + 2\sqrt{3}$

b. $9\sqrt{3} - 5\sqrt{3}$

c. $\sqrt{8} \cdot \sqrt{6}$

d. $\dfrac{12}{\sqrt{3}}$

## Section 10
# *THE PYTHAGOREAN THEOREM*

123. Write the Pythagorean Theorem.

124. Using the Pythagorean Theorem involves squaring expressions and simplifying square roots.  Simplify each expression below as much as you can.

    a. $\sqrt{9^2+12^2}$      b. $\sqrt{4^2+6^2}$      c. $\sqrt{\left(\sqrt{3}\right)^2+\left(\sqrt{2}\right)^2}$

125. Solve each equation.

    a. $x^2=3^2+4^2$      b. $x^2+2^2=10^2$      c. $x^2+\left(4\sqrt{2}\right)^2=8^2$

126. Without using a ruler (the figures are NOT drawn to scale), determine the missing side length for each triangle shown.

a.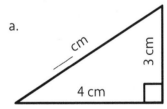
b.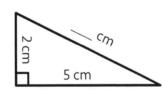
c.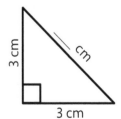

127. ★Determine the missing side length for the triangle shown.

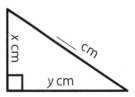

128. Determine the missing side length in each triangle. Write the length in two forms: 1) in simplest radical form, and 2) rounded to the nearest tenth.

a.

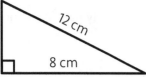

b.

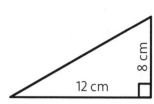

129. The Pythagorean Theorem has a simple structure of $a^2 + b^2 = c^2$ but its simplicity can cause confusion. If the side lengths are labeled $a$, $b$, and $c$, the longest side must be labeled as $c$. The other two side lengths, known as the <u>legs</u> of the right triangle, are interchangeable as $a$ and $b$ or $b$ and $a$. Explain why it doesn't matter which leg is $a$ and which is $b$.

130. Determine the missing side length for each triangle shown.

a.

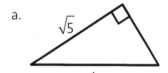

b.

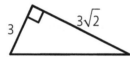

c.

131. ★Draw 3 unique right triangles such that each triangle has a hypotenuse of $\sqrt{10}$. Don't worry too much about drawing them to scale. Just figure out what the side lengths need to be.

132. ★Draw an isosceles right triangle with a hypotenuse of $3\sqrt{2}$. What are the other two lengths?

133. Do you remember how to solve equations when one of the expressions is $x^2$? Bring your mind back to this topic by solving the equations below.

    a. $x^2+6x+8=0$         b. $x^2+6x=0$         c. $3x^2-5x-8=0$

134. ★Find all possible solutions to the equation $6x^2-15x+2=-7$.

135. A ladder leans against a wall. The distance from the top of the ladder to the ground is 3 more feet than the distance from the bottom of the ladder to the wall. The ladder is 15 feet long. Determine the distance from the top of the ladder to the ground.

Section 11

# REVIEW OF RADICAL EXPRESSIONS

136. Multiply each group of expressions and simplify the result as much as you can. Write your answer in the form $A\sqrt{B}$, if possible.

a. $\left(\sqrt{7}\right)\left(-5\sqrt{7}\right)$

b. $\left(\sqrt{3}\right)\left(-4\sqrt{6}\right)$

c. $\left(-2\sqrt{11}\right)^2$

137. Simplify the following expressions.

a. $\left(2+\sqrt{5}\right)\left(2-\sqrt{5}\right)$

b. $\left(4-\sqrt{6}\right)^2$

138. Simplify each expression as much as possible.

a. $\sqrt{\dfrac{8}{18}}$

b. $\dfrac{\sqrt{5}}{\sqrt{25}}$

c. $\dfrac{8\sqrt{7}}{\sqrt{2}}$

139. ★Determine the value of $x$ that makes the equation true.

$$\sqrt{x^2+10}=x+5$$

140. Determine the missing side length for the triangle shown.

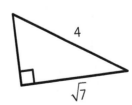

141. Solve the equation $x^2 + 6x = 0$.

142. Which integers have a square root that is less than 5 but greater than 4?

143. How many integers are between the locations of $\sqrt{8}$ and $\sqrt{101}$ on the number line?

144. Which integer is closest to the value of $\sqrt{28} + \sqrt{60}$ ?

Section 12
# CUMULATIVE REVIEW

145. Simplify each expression using only positive exponents in your final answer.

    a. $y^{-3} \cdot y^{-1}$
    b. $2z^{-5}$
    c. $\dfrac{6x^2}{x^{-4}}$

146. Simplify each expression using only positive exponents in your final answer.

    a. $\left(3^{-1}\right)^2$
    b. $\left(f^8\right)^{-2}$
    c. $\left(10g\right)^{-2}$

147. Simplify each expression using only positive exponents in your final answer.

    a. $\left(-5x^{-1}\right)^2$
    b. $\left(-\dfrac{y}{4}\right)^{-2}$

148. Consider the line shown.

    a. Find the equation that shows the relationship between $E$ and $t$.

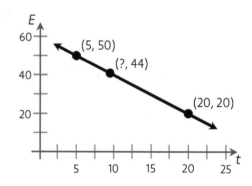

    b. Determine the missing value in the graph.

149. Mary had scores of 74, 81, 99 and 92 on her first four algebra tests. If every test is equally weighted, what must she score on the next test to obtain an average of at least. . .

    a. 87%?
    b. 90%?

150. Find the *x*-intercept and the *y*-intercept of the line given by $-2x + 4y = 13$. Find the coordinates of two more ordered pairs and then graph the line.

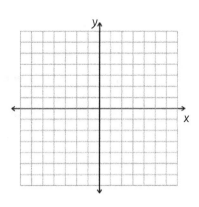

151. Evaluate each expression if *x* is assigned the value of $-2$. Calculators are NOT allowed here.

    a. $-\dfrac{1}{2}x$          b. $5x^2$          c. $-5x^3$          d. $\left(-x\right)^4$          e. $\dfrac{1}{16}x^5$

152. When a valve is opened, water flows through the valve to fill a swimming pool at a rate of 25 gallons per minute. If a swimming pool already holds 8,500 gallons of water, how many hours will it take to fill the pool to its capacity of 20,000 gallons?

153. A number is decreased by 40% and this new number is raised by 25%. The result is 225. What was the original number?

154. ★Determine the value of *x* that will make the shaded region have an area of 20 units$^2$.

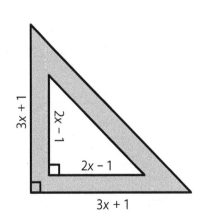

## Section 13
# ANSWER KEY

| | |
|---|---|
| 1. | a. $\dfrac{3}{2}$    b. $\dfrac{35}{12}$    c. $\dfrac{8}{27}$    d. $-\dfrac{19}{12}$ |
| 2. | $y = mx + b$ |
| 3. | $Ax + By = C$ |
| 4. | $y = -\dfrac{3}{4}x$ |
| 5. | $4x - 3y = -15 \rightarrow y = \dfrac{4}{3}x + 5$ |
| 6. | $(-2.4, 1.8)$ |
| 7. | $2{:}\ 4, 8, 16$    $(-3){:}\ 9, -27, 81$    $4{:}\ 16, 64, 256$ |
| 8. | $(-5){:}\ 25, -125, 625$    $6{:}\ 36, 216, 1296$ <br> $(-7){:}\ 49, -343, 2401$ |
| 9. | a. $-32$    b. $81$    c. $-512$ |
| 10. | a. $7$    b. $5$    c. $4$    d. $2$ |
| 11. | a. $11$    b. $3$    c. $2$ |
| 12. | a. $1$    b. $0$ |
| 13. | a. $-6$    b. $-10$ |
| 14. | a, b. not possible |
| 15. | You can check that your square root is accurate by squaring your result. Based on what you have learned so far, it is not possible for a number to be negative <u>after</u> it has been squared. Similarly, the fourth root of a negative number is not possible because a nonzero number raised to the fourth power is positive, not negative. |
| 16. | a. $1$    b. $-3$    c. $5$    d. $-2$ |
| 17. | a. $-10$    b. $6$    c. $-7$    d. not possible |
| 18. | $(-5)^2 = 25$. A negative number multiplied by itself is positive. |
| 19. | a. $\dfrac{8}{27}$    b. $\dfrac{1}{49}$    c. $-\dfrac{125}{8}$ |
| 20. | a. $\dfrac{x^4}{16}$    b. $\dfrac{9y^2}{16}$    c. $-\dfrac{32}{x^{10}y^5}$ |
| 21. | $\dfrac{1}{2} \rightarrow \left(\dfrac{1}{2}\right)^4 = \dfrac{1}{16}$ |
| 22. | a. $\dfrac{1}{5}$    b. $-\dfrac{3}{4}$    c. $-\dfrac{3}{10}$    d. impossible |
| 23. | You can check that your square root is |

| | |
|---|---|
| | accurate by squaring your result. Based on what you have learned so far, it is not possible for a number to be negative <u>after</u> it has been squared. |
| 24. | a. $\dfrac{3y}{4}$    b. $\dfrac{x}{2}$    c. $-\dfrac{3x}{10y^2}$ |
| 25. | It is not possible to find this value without a calculator. |
| 26. | There are seven: 1, 8, 27, 64, 125, 216, 343 |
| 27. | $1000 - 10 = 990$ |
| 28. | a. 2 out of 8; $\dfrac{1}{4}(25\%)$ <br><br> b. 5 out of 125; $\dfrac{1}{25}(4\%)$ <br><br> c. 10 out of 1,000; $\dfrac{1}{100}(1\%)$ |
| 29. | a. slightly higher than 2 (2.1) <br> b. slightly higher than 4 (4.1) <br> c. slightly lower than 5 (4.9) |
| 30. | a. 2.08    b. 4.12    c. 4.93 |
| 31. | 990. There are 10 perfect cubes from 1 to 1,000. The cube root of a perfect cube is rational. The remaining 990 integers have irrational cube roots. |
| 32. | There are nine: 1, 4, 9, 16, 25, 36, 49, 64, 81 |
| 33. | a. 3.3    b. 8.8    c. 7.1 |
| 34. | |
| 35. | a. 30    c. 14    e. 28 <br> b. 30    d. 14    f. 28 |
| 36. | a. $\sqrt{A}\cdot\sqrt{B}$    b. $\sqrt{X\cdot Y}$ |
| 37. | $10\sqrt{3}$ |
| 38. | a. four root two    b. two root five <br> c. nine root seven |
| 39. | a. $4\sqrt{2}$    b. $2\sqrt{5}$ |
| 40. | a. $9\sqrt{7}$    b. $\sqrt{33}$ |
| 41. | 1, 4, 9, 16, 25, 36, 49, 64, 81, 100, 121, 144 |
| 42. | $2\sqrt{3}$ |
| 43. | a. $3\sqrt{6}$    b. $\sqrt{16}\sqrt{5} \rightarrow 4\sqrt{5}$ |

| | |
|---|---|
| | c. $\sqrt{4}\sqrt{13} \to 2\sqrt{13}$ |
| 44. | a. $2\sqrt{6}$   b. $2\sqrt{10}$   c. $\sqrt{30}$ |
| 45. | a. 25   b. 64 |
| 46. | a. 100   b. 36 |
| 47. | a. $2\sqrt{7}$   b. $4\sqrt{2}$ |
| 48. | a. $2\sqrt{15}$   b. $\sqrt{29}$ |
| 49. | $27 \to \sqrt{27} = \sqrt{9}\sqrt{3} = 3\sqrt{3}$ |
| 50. | $B^2$ |
| 51. | a. 45   b. 88   c. 250 |
| 52. | a. 18   b. $3 \cdot 2\sqrt{5} \to 6\sqrt{5}$<br>c. $2 \cdot 3\sqrt{3} \to 6\sqrt{3}$ |
| 53. | a. $7\sqrt{4 \cdot 2} \to 7 \cdot 2\sqrt{2} \to 14\sqrt{2}$<br>b. $5\sqrt{36 \cdot 2} \to 5 \cdot 6\sqrt{2} \to 30\sqrt{2}$<br>c. fully simplified |
| 54. | a. $y = 4$   b. $y = 49$   c. $y = 50$<br>d. not possible |
| 55. | $M^2$ |
| 56. | $T^2 - 5$ |
| 57. | a. $x = 19$   b. $x = 2$   c. not possible |
| 58. | not possible |
| 59. | a. $x = -1$   b. $x = 1$   c. $x = -12$ |
| 60. | a. $x = 3$   b. $x = 18$   c. $x = 4$ |
| 61. | a. $x^2 - 16x + 64$   b. $25 - 20x + 4x^2$ |
| 62. | a. $(x-4)(x-3) = 0 \to x = 4$ or 3<br>b. $(4x+3)(x-5) = 0 \to x = -\frac{3}{4}$ or 5 |
| 63. | $4x^2 - 3x = 4x^2 - 4x + 1 \to x = 1$ |
| 64. | a. $x = 8$ ($x = 2$ or 8, but 2 doesn't work)<br>b. $x = 11$ ($x = 11$ or 4, but 4 doesn't work) |
| 65. | – |
| 66. | $x = 5$ |
| 67. | Solve $\sqrt{N-7} = 3 : N = 16$ |
| 68. | Solve $\sqrt{x+1} = 0.5 : x = -0.75$ |
| 69. | 8 cm (each side is 2 cm) |
| 70. | Solve $\sqrt[3]{K-5} = -2 : K = -3$ |
| 71. | $K = 20, L = 45, M = 80$   $K+L+M = 145$ |
| 72. | $5\sqrt{3},\ 4\sqrt{5},\ 3\sqrt{9},\ 2\sqrt{21}$<br>$\left(\sqrt{75}, \sqrt{80}, \sqrt{81}, \sqrt{84}\right)$ |
| 73. | The largest perfect square factor of 54 is 9. Since $54 = 9 \cdot 6$, then $\sqrt{54} = \sqrt{9} \cdot \sqrt{6}$, which can be simplified as $3\sqrt{6}$. |
| 74. | a. $20\sqrt{2}$   b. $3\sqrt{5}$ |
| 75. | a. 2   b. $6\sqrt{10}$   c. 6 |

| | |
|---|---|
| 76. | a. $20y$   b. $x^2\sqrt{y}$   c. $x$ |
| 77. | a. $6\sqrt{2}$   b. $18\sqrt{7}$   c. 18 |
| 78. | a. $16xy$   b. $8\sqrt{14}$   c. 20   d. $x^2 y$ |
| 79. | a. $30\sqrt{3}$   b. $-60$   c. $-9\sqrt{2}$   d. 24 |
| 80. | $\sqrt{12} = 2\sqrt{3}$ ; $\sqrt{48} = 4\sqrt{3} = 2(3.46) = 6.92$ |
| 81. | a. $y = -12$   b. $f = 11$   c. $x = 37$ |
| 82. | a. $5\sqrt{5}$   c. $\sqrt{5}$   e. 6<br>b. 15   d. 1   f. $2\sqrt{3} + 2\sqrt{6}$ |
| 83. | No. $\sqrt{4} + \sqrt{4} = 2 + 2 = 4$ ; $\sqrt{4+4} = \sqrt{8} = 2\sqrt{2}$ |
| 84. | a. No.   b. No. |
| 85. | a. $2\sqrt{5}$   b. $2\sqrt{7}$   c. $4\sqrt{11}$ |
| 86. | a. $3\sqrt{3}$   b. Fully simplified<br>c. $8\sqrt{2}$   d. Fully simplified |
| 87. | Terms are "like terms" if they have the same value inside the radical symbol. |
| 88. | a. $2\sqrt{6}$   b. $\sqrt{5}$ |
| 89. | $3\sqrt{11} + 2\sqrt{11} = 5\sqrt{11}$ |
| 90. | a. $-2\sqrt{7} + 2\sqrt{7} = 0$   b. $2\sqrt{2} - 2\sqrt{3}$ |
| 91. | a. $2\sqrt{3} + \sqrt{3} = 3\sqrt{3}$<br>b. $4\sqrt{2} - 2\sqrt{2} = 2\sqrt{2}$ |
| 92. | a. $2\sqrt{5} + 2\sqrt{6}$   b. $3\sqrt{7} - 2\sqrt{7} = \sqrt{7}$ |
| 93. | a. $2\sqrt{90} + 2\sqrt{160} \to 2\sqrt{9}\sqrt{10} + 2\sqrt{16}\sqrt{10}$<br>$2 \cdot 3\sqrt{10} + 2 \cdot 4\sqrt{10} \to 14\sqrt{10}$ cm<br>b. $\sqrt{90} \cdot \sqrt{160} \to 3\sqrt{10} \cdot 4\sqrt{10} \to 12\sqrt{100}$<br>$\to 12 \cdot 10 \to 120$ cm$^2$ |
| 94. | 0 |
| 95. | a. $5w - 30$   b. $-2x^3 + x$<br>c. $-3y^5 - 24y^4 + 30y^2$ |
| 96. | a. $6 - 2\sqrt{5}$   b. $2\sqrt{6} - 3\sqrt{2}$   c. 15 |
| 97. | a. $2x^2 - 4x + 7x - 14 \to 2x^2 + 3x - 14$<br>b. $4x^2 - 36x + 81$   c. $x^2 - 9y^2$ |
| 98. | a. 14   b. 3   c. $-7$ |
| 99. | a. $18 - 8\sqrt{2}$   b. $15 - 6\sqrt{6}$   c. $9 + 4\sqrt{2}$ |
| 100. | When you multiply binomials with irrational expressions, if two of the resulting terms are irrational expressions with opposite signs, they cancel each other because they make zero. |
| 101. | a. 5   b. $38 + 12\sqrt{2}$   c. 46 |
| 102. | a. 1   b. $-65$   c. $2 + 12\sqrt{2} - 2\sqrt{6} - 24\sqrt{3}$ |
| 103. | a. $\sqrt{3}$   b. $\sqrt{5}$   c. $7 + \sqrt{2}$   d. $3 - 5\sqrt{7}$ |

| # | Answer |
|---|--------|
| 104. | $\sqrt{8}$ or $2\sqrt{2}$ inches |
| 105. | a. $\dfrac{1}{6}$  b. $\dfrac{4}{7}$  c. $\dfrac{9}{2}$  d. $\dfrac{11}{12}$ |
| 106. | a. 1  b. $\sqrt{3}$  c. 3  d. 6 |
| 107. | a. 2  b. $\sqrt{5}$  c. 2  d. $\sqrt{5}$ |
| 108. | a. $\dfrac{9}{4}$  b. $\sqrt{2}$  c. $\dfrac{9}{4}$  d. $\sqrt{2}$ |
| 109. | a. $\dfrac{\sqrt{x}}{\sqrt{y}}$  b. $\sqrt{\dfrac{f}{g}}$ |
| 110. | a. $\dfrac{3}{12}\to\dfrac{1}{4}$  b. $\dfrac{11}{4}$  c. $\dfrac{10}{6}\to\dfrac{5}{3}$ |
| 111. | a. $\sqrt{\dfrac{1}{9}}=\dfrac{1}{3}$  b. $\sqrt{4}=2$  c. $\sqrt{5}$ |
| 112. | a. $\dfrac{\sqrt{3}}{4}$  b. $\dfrac{\sqrt{6}}{5}$  c. $\dfrac{2\cdot3\sqrt{2}}{4\sqrt{2}}\to\dfrac{6}{4}\to\dfrac{3}{2}$ |
| 113. | a. $\dfrac{2}{\sqrt{3}}$  b. $\dfrac{1}{2\sqrt{3}}$  c. $\dfrac{5\sqrt{2}}{\sqrt{7}}$ |
| 114. | There is a radical expression in the denominators in the second of the previous two scenarios. |
| 115. | a. $\dfrac{\sqrt{3}}{\sqrt{3}}$  b. $\dfrac{\sqrt{5}}{\sqrt{5}}$  c & d. already rational |
| 116. | a. $\sqrt{2}$  b. $\dfrac{\sqrt{3}}{2}$  c. $\sqrt{3}$  d. $3\sqrt{10}$ |
| 117. | a. $\dfrac{5\sqrt{2}}{4}$  b. $\dfrac{5\sqrt{2}}{4}$  c. both yield the same result |
| 118. | a. $\dfrac{\sqrt{3}}{3}$  b. $\dfrac{3\sqrt{2}}{2}$  c. $\dfrac{\sqrt{15}}{9}$ |
| 119. | a. $\sqrt{5}$  b. $\sqrt{2}$  c. $\dfrac{6\sqrt{3}}{5}$ |
| 120. | $\dfrac{\sqrt{2}+\sqrt{6}}{2}$ |
| 121. | Each expression simplifies to $\dfrac{6}{7}$. |
| 122. | Each expression simplifies to $4\sqrt{3}$. |
| 123. | $a^2+b^2=c^2$ |
| 124. | a. 15  b. $2\sqrt{13}$  c. $\sqrt{5}$ |
| 125. | a. 5  b. $4\sqrt{6}$  c. $4\sqrt{2}$ |
| 126. | a. 5  b. $\sqrt{29}$  c. $3\sqrt{2}$ |
| 127. | $\sqrt{x^2+y^2}$ |
| 128. | a. $4\sqrt{5}\approx8.9$  b. $4\sqrt{13}\approx14.4$ |
| 129. | The commutative property of addition: $a+b=b+a$ |
| 130. | a. $\sqrt{11}$  b. $3\sqrt{3}$  c. 4 |
| 131. | Possible side lengths: $1,3,\sqrt{10}$; $\sqrt{2},2\sqrt{2},\sqrt{10}$; $\sqrt{3},\sqrt{7},\sqrt{10}$; $2,\sqrt{6},\sqrt{10}$; $\sqrt{5},\sqrt{5},\sqrt{10}$ |
| 132. | side $a$ and side $b$ must be 3 |
| 133. | a. $(x+2)(x+4)=0\to x=-2,-4$  b. $x(x+6)=0\to x=0,-6$  c. $(3x-8)(x+1)=0\to x=\dfrac{8}{3},-1$ |
| 134. | $6x^2-15x+9=0\to3(2x^2-5x+3)=0$  $3(2x-3)(x-1)=0\to x=\dfrac{3}{2},1$ |
| 135. | Solve $x^2+(x+3)^2=15^2\to12$ feet |
| 136. | a. $-35$  b. $-12\sqrt{2}$  c. 44 |
| 137. | a. $-1$  b. $22-8\sqrt{6}$ |
| 138. | a. $\dfrac{2}{3}$  b. $\dfrac{\sqrt{5}}{5}$  c. $4\sqrt{14}$ |
| 139. | $x=-\dfrac{3}{2}$ |
| 140. | 3; solve $x^2+(\sqrt{7})^2=4^2$ |
| 141. | $x=0,-6$ |
| 142. | 17, 18, 19, 20, 21, 22, 23, 24 |
| 143. | 8 total (3, 4, 5, 6, 7, 8, 9, 10) |
| 144. | approx. 5 + approx. 8 = approx. 13 |
| 145. | a. $\dfrac{1}{y^4}$  b. $\dfrac{2}{z^5}$  c. $6x^6$ |
| 146. | a. $\dfrac{1}{9}$  b. $\dfrac{1}{f^{16}}$  c. $\dfrac{1}{100g^2}$ |
| 147. | a. $\dfrac{25}{x^2}$  b. $\dfrac{16}{y^2}$ |
| 148. | a. $E=-2t+60$  b. 8 |
| 149. | a. $\dfrac{74+81+99+92+x}{5}=87\to x=89$  b. not possible, unless you can earn a score of 104% |
| 150. | $x$-int: $(-6.5, 0)$; $y$-int: $(0, 3.25)$ |

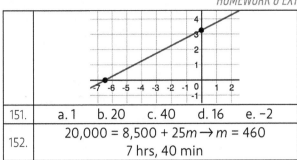

| | | | | | |
|---|---|---|---|---|---|
| 151. | a. 1 | b. 20 | c. 40 | d. 16 | e. −2 |

| 152. | $20{,}000 = 8{,}500 + 25m \rightarrow m = 460$ <br> 7 hrs, 40 min |
|---|---|

| 153. | $(.6x) + .25(.6x) = 225 \rightarrow x = 300$ |
|---|---|
| 154. | $\dfrac{1}{2}(3x+1)^2 - \dfrac{1}{2}(2x-1)^2 = 20 \rightarrow x = 2$ |

CPSIA information can be obtained
at www.ICGtesting.com
Printed in the USA
LVHW051716140820
663222LV00008B/421